Advancing Maths for AQA

MECHANICS I

Ted Graham, Aidan Burrows and Joan Corbett

Series editors
Ted Graham Sam Boardman Graham Eaton
Keith Parramore Roger Williamson

Heinemann

Heinemann Educational Publishers
a division of Heinemann Publishers (Oxford) Ltd,
Halley Court, Jordan Hill, Oxford OX2 8EJ

OXFORD MELBOURNE AUCKLAND JOHANNESBURG
BLANTYRE GABORONE PORTSMOUTH NH (USA) CHICAGO

First published in 2000

01 10 9 8 7 6 5 4 3 2 1

ISBN 0 435 51306 0

Original design by Wendi Watson

Cover design by Miller, Craig and Cocking

Typeset and illustrated by Tech-Set Limited, Gateshead, Tyne & Wear

Printed and bound by Bath Press in the UK

Acknowledgements
The publishers and authors acknowledge the work of the writers, Ray Atkin, John Berry, Derek Collins, Tim Cross, Ted Graham, Phil Rawlins, Tom Roper, Rob Summerson, Nigel Price, Frank Chorlton and Andy Martin of the *AEB Mathematics for AS and A-Level* Series, from which some exercises and examples have been taken.

The publishers' and authors' thanks are due to the AEB for permission to reproduce questions from past examination papers.

The answers have been provided by the authors and are not the responsibility of the examining board.

About this book

This book is one in a series of textbooks designed to provide you with exceptional preparation for AQA's new Advanced GCE Specification B. The series authors are all senior members of the examining team and have prepared the textbooks specifically to support you in studying this course.

Finding your way around

The following are there to help you find your way around when you are studying and revising:

- **edge marks** (shown on the front page) – these help you to get to the right chapter quickly;
- **contents list** – this identifies the individual sections dealing with key syllabus concepts so that you can go straight to the areas that you are looking for;
- **index** – a number in bold type indicates where to find the main entry for that topic.

Key points

Key points are not only summarised at the end of each chapter but are also boxed and highlighted within the text like this:

> For every action, there is an equal but opposite reaction

Exercises and exam questions

Worked examples and carefully graded questions familiarise you with the specification and bring you up to exam standard. Each book contains:

- Worked examples and Worked exam questions to show you how to tackle typical questions; Examiner's tips will also provide guidance;
- Graded exercises, gradually increasing in difficulty up to exam-level questions, which are marked by an [A];
- Test-yourself sections for each chapter so that you can check your understanding of the key aspects of that chapter and identify any sections that you should review;
- Answers to the questions are included at the end of the book.

Contents

CHAPTER 1

Mathematical modelling in mechanics

Learning objectives

After studying this chapter you should:

- be aware of the mathematical modelling cycle
- understand the types of assumptions used when modelling problems
- be aware of the difference between particle and rigid body models
- be familiar with some of the terminology used in mechanics.

1.1 Introducing modelling

Mathematical modelling describes the process of obtaining a solution to a real world problem. Many real problems can be very complex and so the idea of creating a mathematical model is to simplify the real situation, so that it can be described using equations or graphs. These equations or graphs are referred to as a mathematical model. These mathematical models can provide solutions to the original problem. It is often necessary to interpret these answers in the context of the original problem and to check that the answers that you have obtained are reasonable.

Examples of problems where modelling could be used:

- to determine the maximum speed of a car round a bend
- to help define the design requirements of a sports stadium
- to evaluate new design options for a mountain bike
- to work out how to send a space station into orbit.

> This whole process is often referred to as the mathematical modelling cycle because it may be necessary to repeat the process, creating better models of reality until a satisfactory solution is obtained.

The key stages of the mathematical modelling cycle are shown in the diagram on the next page.

The mathematical modelling cycle

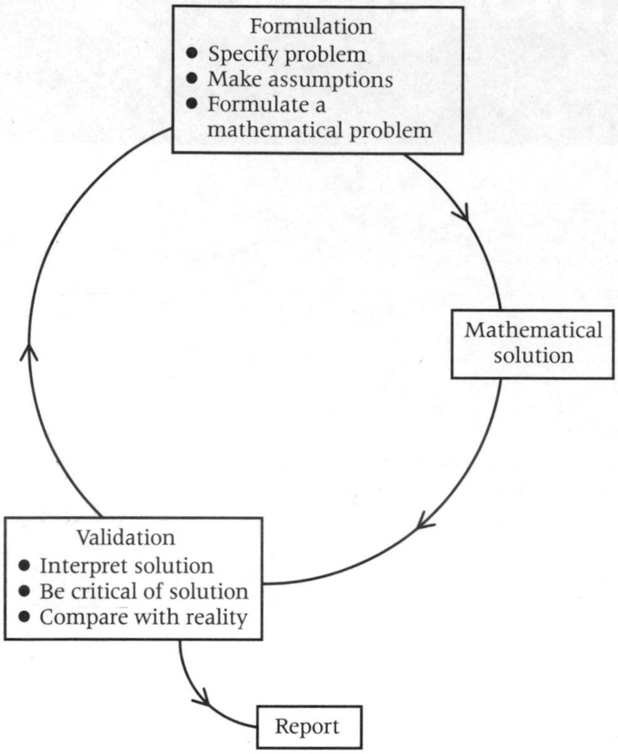
Formulation
• Specify problem
• Make assumptions
• Formulate a mathematical problem
Mathematical solution
Validation
• Interpret solution
• Be critical of solution
• Compare with reality
Report

Each stage of this cycle is now considered in more detail.

1.2 The modelling cycle

Formulation

This stage of the modelling cycle consists of three distinct activities:

- specifying the problem
- making assumptions
- formulating a mathematic problem.

Specify problem

This is the starting point of the modelling cycle. Very often these problems will be very open-ended, unlike the types of problems that you will find in textbook exercises. Your first task is to make sure that you understand the problem and have decided what you need to find to solve the problem.

Make assumptions

The next part of the formulation stage is to make assumptions to simplify the problem. Some of the assumptions that you will often make in mechanics are now considered, in the context of a body that remains at rest or moves in some way.

Assume that the body under consideration is a **particle**. This means that you assume that the body has no size, but does have a mass. So any rotation of the body is ignored, any forces will all act in the one place and the position is precisely defined. It may be quite reasonable to model a car as a particle if it is being driven a large distance or to model a ball as a particle.

A more sophisticated model is that of a **rigid body**. Here the body is assumed to have size, but to be rigid, so that it does not compress or change shape. The simplest rigid body is a rod. With a rigid body the forces will probably not all act at the same place and this must be taken into account. If you want to model the motion of a rolling ball it will be important to consider its size and shape, so a rigid sphere may be a suitable body.

A particle has no size but does have a mass.

A rigid body has size but does not change shape when forces are applied to it.

There are some factors, such as air resistance, lift on a golf ball, friction, resistance in a pulley, that may make a problem hard to describe and solve mathematically. We often ignore these factors in order to develop a simple mathematical model, but may have to incorporate them at a later stage if our results do not give good predictions. Further examples of things that we might ignore are the mass of a string, friction, that a string might be elastic and stretch.

Any assumptions made should be clearly recorded so that they can be reviewed later in the process.

Some keywords that you will find have particular meanings in a modelling context are listed below:

• smooth	no friction
• rough	friction present
• light	has no mass
• inelastic	does not stretch
• inextensible	does not stretch.

Formulate a mathematical problem

In this phase of the modelling cycle you turn the open-ended problem that you started with into a focused mathematical problem that you can solve using mathematical techniques.

You may also need to collect some data that is relevant to the problem that you have to solve.

You may also need to introduce algebraic variables to represent physical quantities that are important in the context of the problem. For example:

m the mass of the body
u the initial velocity of the body,
etc.

Finally in the formulation stage there are a number of standard mathematical models to describe physical phenomena, for example to calculate the friction present, the size of the gravitational attraction, the force exerted by a stretched spring. You will learn about these models as you work through the mechanics modules and be able to include them in any modelling that you do.

Mathematical solution

This is where you obtain an actual solution, by carrying out calculations, solving equations and using other mathematical techniques. As you learn more pure mathematics you will be able to deal with more sophisticated mathematical models. When you have obtained a solution you move on to the interpretation and validation phase.

Validation

This is where you interpret the mathematical answers in the context of the original problem. This is where you will probably reach a conclusion of some kind. For example you may state that it is not safe for a lorry to drive over a bridge, that a car was breaking a speed limit or that it is impossible to hit a tennis ball over the net from a certain position.

It is also important to validate your answer, to check that it is reasonable. This may require you to make some observations of reality or to carry out an experiment to validate your predictions or it may simply be that the model you have created produced totally unrealistic solutions. For example if you calculate that a cyclist rolling down a hill will reach a speed of 70 mph after travelling 100 m, by ignoring air resistance, then it is clear that you need to go back to your original assumptions and try to include air resistance.

If you do need to go back to your original assumptions and reformulate your model, then you are embarking on a second cycle, which should lead to a more realistic solution. Finally when you obtain a satisfactory solution, you will need to report on your conclusion, describing how you reached this conclusion.

Worked example 1

How far apart should speed bumps be placed so that traffic does not reach a speed greater than 30 mph?

Solution

Assume that:

- A car is a particle.
- All cars slow down according to the table of stopping distances in the Highway Code.
- All cars speed up at the same rate as they slow down.

Gathering data from the Highway Code gives a figure of 23 m as the stopping distance at 30 mph.

Based on our assumptions we formulate the mathematical model:

Distance between speed bumps = 2 × stopping distance

Using this model we can calculate the required distance as 46 m (the distance to speed up to 30 mph and to slow down from 30 mph).

A person solving this problem in real life may be able to set up some experimental bumps and observe the speeds of cars between them.

There are some ways in which this model could be revised or reformulated, for example:

- Modelling the car as a rigid body that has length.
- Looking for alternative models to describe how a car gains speed and slows down.

EXERCISE 1A

1 Gather some data on the lengths of cars and revise the solution to the speed bumps problem to take account of this factor.

2 Make a list of the assumptions that you might make to model the motion of a javelin.

3 If you were to model the motion of the pendulum in a grandfather clock:

(a) make a list of the assumptions that you might make

(b) make a list of the variables that you might include in your model.

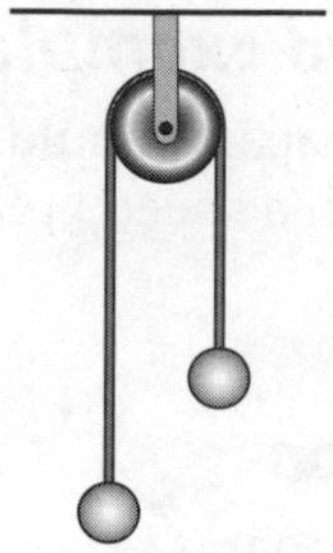

4 A string that passes over a pulley connects two objects of equal masses. The initial positions of the objects are shown in the diagram. Discuss the different predictions that you would obtain if you model:

(a) the pulley as smooth

(b) the pulley as smooth and the string as light.

5 A student models a parachutist as a particle that does not experience air resistance. Suggest what predictions he might obtain.

Key point summary

1 The mathematical modelling cycle consists of: *p2*

- formulating the problem
- obtaining a mathematical solution
- validating and interpreting the solution
- preparing a report.

2 Modelling requires assumptions to be made. *p2*

3 Particle and rigid body models are different as a particle has no size but has a mass, whereas a rigid body has size but does not change shape. *p3*

4 Terminology used in mechanics: *p3*

• Smooth	no friction
• Rough	friction present
• Light	has no mass
• Inelastic	does not stretch
• Inextensible	does not stretch.

The mathematical modelling cycle

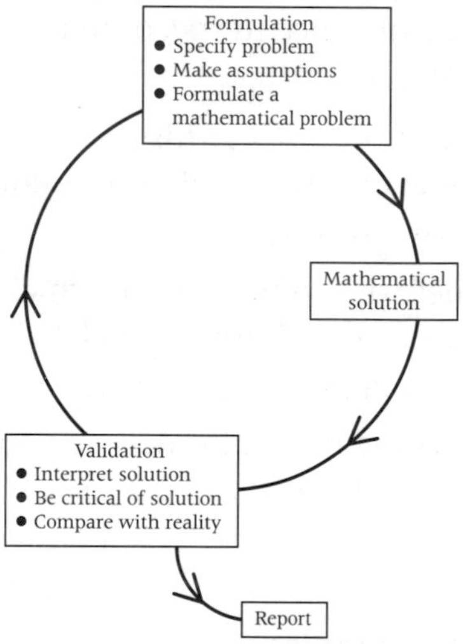
Formulation
• Specify problem
• Make assumptions
• Formulate a mathematical problem
Mathematical solution
Validation
• Interpret solution
• Be critical of solution
• Compare with reality
Report

CHAPTER 2

Kinematics in one dimension

Learning objectives

After studying this chapter you should be able to:

- define displacement, velocity and acceleration
- understand and interpret displacement–time graphs and velocity–time graphs
- solve problems involving motion under constant acceleration using standard formulae
- solve problems involving motion under gravity.

2.1 Introducing kinematics

Kinematics is the study of motion and in this chapter we will study objects which move in one dimension only, this means that they move in a straight line. Examples are:

- cars, buses, or bikes, etc. on a straight road
- objects dropped from the top of a cliff or tower
- an athlete running in a 100 m race.

We will not, at this stage, consider what makes the objects move!

We shall also model most of the objects as particles so that we can ignore their size and shape and concentrate on their movement.

2.2 Displacement, velocity and acceleration

First, we have to define some words and symbols that we will use. In everyday language we use the terms **speed** and **distance** when talking about motion. **Distance** is how far we travel (miles, metres, etc.) and **speed** is how fast we go (miles per hour, metres per second, etc.). In the study of kinematics we need to be more precise and so we introduce **displacement** and **velocity**.

Displacement is based on the distance from a specific origin or reference point, but it also takes account of the direction in which the particle has moved.

The displacement may be 5 km north from a reference point or origin; or it may be given using positive or negative values relative to an origin, as shown in the diagram.

We decide that, relative to the point O, displacements to the right are positive and those to the left negative. Thus the displacement of P is $+3$ cm, and of Q -2 cm.

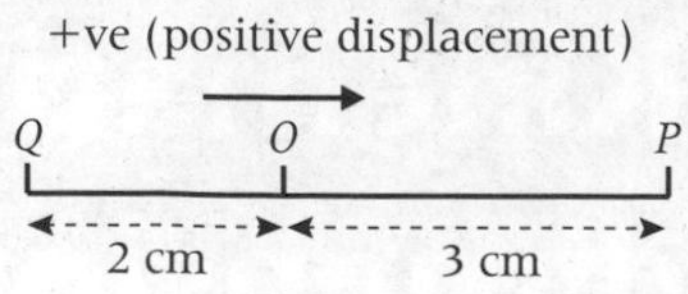

In mechanics the symbol used for displacement is s.

Velocity is defined similarly using the speed of an object together with the direction of the motion.

The velocity could be 6 mph going south-west from a reference point, or positive and negative values can be used.

In the diagram two particles are moving vertically. A is going up at 4 m s^{-1} and B is falling at 5 m s^{-1}. We choose upwards, say, as the positive direction so the velocity of A is $+4\text{ m s}^{-1}$ and B is -5 m s^{-1}.

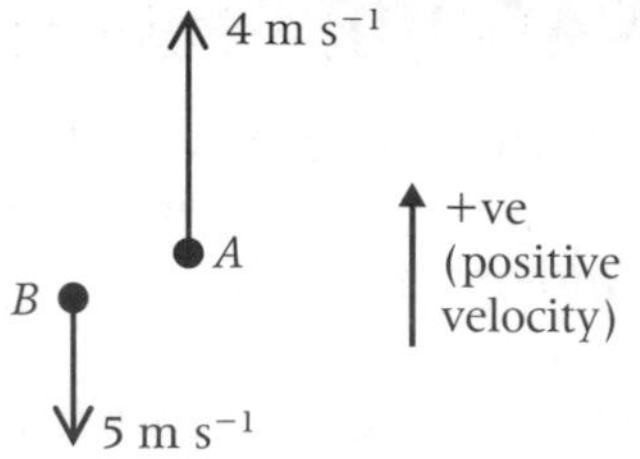

The symbol used for velocity is v.

We often speak of **average speed** meaning the **constant speed** we could have travelled at in order to cover a journey in the same time. For example, if you travel a journey of 100 miles in 2 hours, your average speed is 50 mph; it is very unlikely that you could have driven at a constant 50 mph!

(Quantities that have size (magnitude) but no specific direction, such as speed and distance, are called **scalars**, but those with direction, like velocity and displacement, are called **vectors**. Vectors are studied in more detail in Chapter 3.)

Displacement–time graphs

As a body moves the displacement changes so that s is a function of time, t. A graph plotting s against t, is called a displacement–time graph. As an example the graph shows the displacement–time graph for the motion of a ball thrown in the air and falling back to the floor.

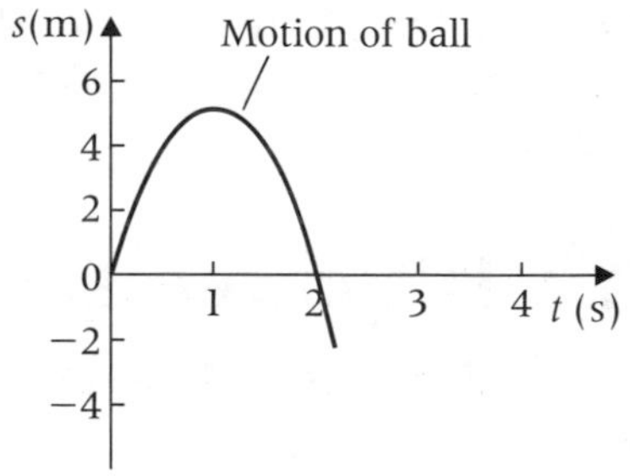

The motion of the ball can be described by looking at the graph. The ball starts at the origin and begins to move in the positive s direction (upwards) to a maximum height of 5 m above the point of release. It then falls to the floor which is 2 m below the point of release. It takes just over 2 seconds to hit the floor.

In another example of a displacement–time graph, the table shows the displacement (s) of a boy, who is running on a straight track, measured at 2-second intervals.

Time, t	0	2	4	6	8	10	12
Displacement, s	0	4	8	12	16	20	24

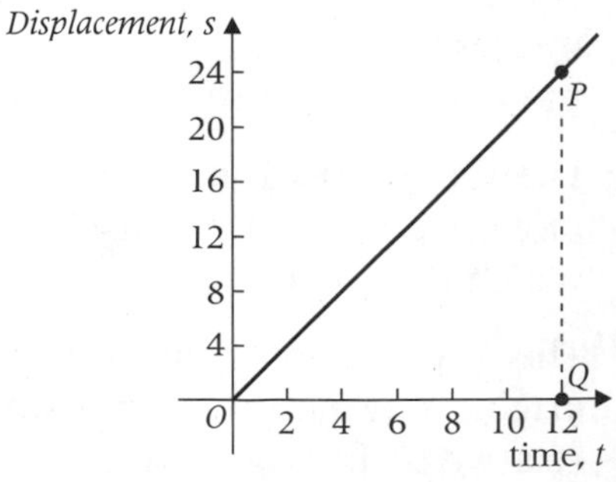

These values have been plotted on the graph. You notice that this graph is a straight line which tells us that the boy is running with a constant velocity.

His velocity is calculated from the gradient of the graph

$$\text{gradient} = \frac{PQ}{OQ} = \frac{24}{12} = 2$$

So the boy runs at 2 m s^{-1}

2

Velocity–time graphs

If we know the velocity, v, at time, t, then we can draw a graph of velocity against time.

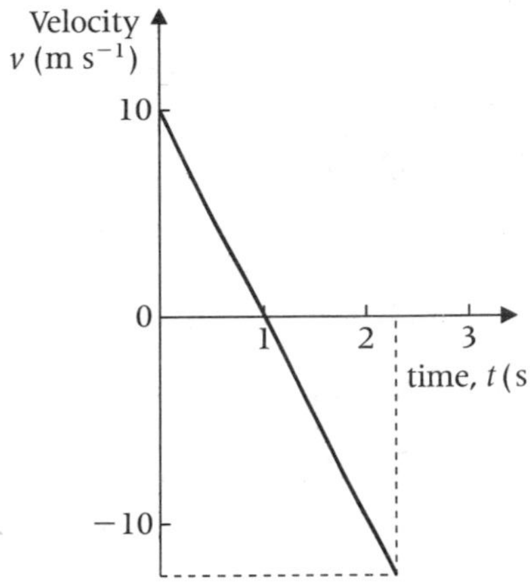

This velocity–time graph is for the motion of the ball whose displacement–time graph we saw above. The ball is thrown up into the air with a velocity of 10 m s^{-1} and this decreases to zero as the ball reaches the highest point, in about 1 second. The direction of motion is then reversed as the ball falls back to the floor, so for this part of the motion the velocity is negative. The ball hits the floor with a velocity of about -12 m s^{-1}

Here is another example of a velocity–time graph:

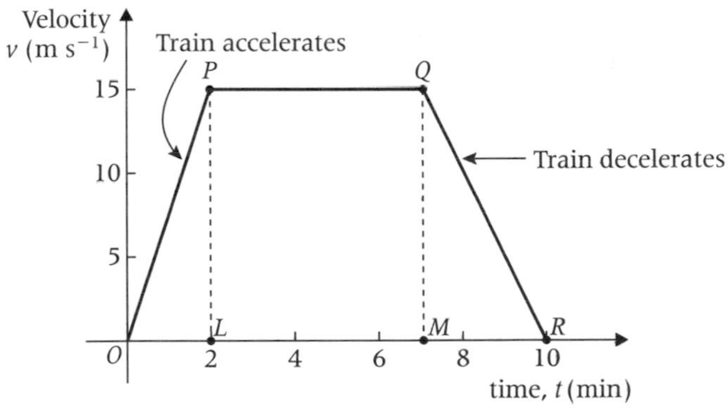

The velocity–time graph here is for the motion of a train which starts from rest at a station. Its velocity increases to 15 m s^{-1} in 2 minutes. Then it travels at that speed for 5 minutes before slowing down over a 3-minute period to stop at the next station.

Acceleration

Another familiar word used to describe the motion of cars, trains, bikes, and so on, in everyday language is **acceleration**. In the last two graphs above the velocity changes and the rate of change is called the acceleration.

Acceleration is defined as the rate at which the velocity is changing. Its units are 'metres per second per second' or m s^{-2}. So an acceleration of 5 m s^{-2} means that the velocity is increasing by 5 m s^{-1} every second.

> We can calculate the acceleration from the gradient of a velocity–time graph. Look at the v–t graph of the train above.

During the first part of the motion:

$$\begin{aligned}\text{accleration} &= \text{gradient of } OP \\ &= \frac{15}{120} \\ &= \frac{1}{8} \text{ or } 0.125\ \text{m s}^{-2}\end{aligned}$$

Note that the time has been converted from minutes to seconds.

During the middle section the acceleration is zero.

In the third section of the train's journey it is slowing down; this is "decelerating" or "retarding" and the acceleration will have a negative value.

$$\begin{aligned}\text{acceleration} &= \text{gradient of } QR \\ &= \frac{-15}{180} \\ &= -\frac{1}{12} \text{ or } -0.0833\ \text{m s}^{-2} \text{ (to three significant figures)}\end{aligned}$$

Note that the symbol used for acceleration is a.

Displacement and the velocity–time graph

If a lorry moves at a constant speed of 15 m s^{-1} for 10 seconds, how far does it travel?
The distance that the lorry travels is simply given by

$$15 \times 10 = 150\ \text{m}$$

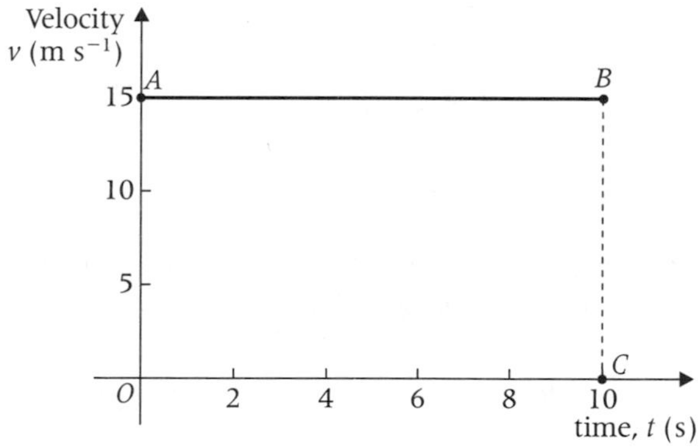

If we look at the velocity–time graph for the lorry, we can see that 150 is the area of the rectangle $OABC$. We refer to this as the "the area under the graph".

This method of calculating distance travelled from a v–t graph can be extended to cases where the velocity is not uniform.

Later the lorry accelerates from rest to 20 m s^{-1} in 8 seconds. The diagram shows its v–t graph.

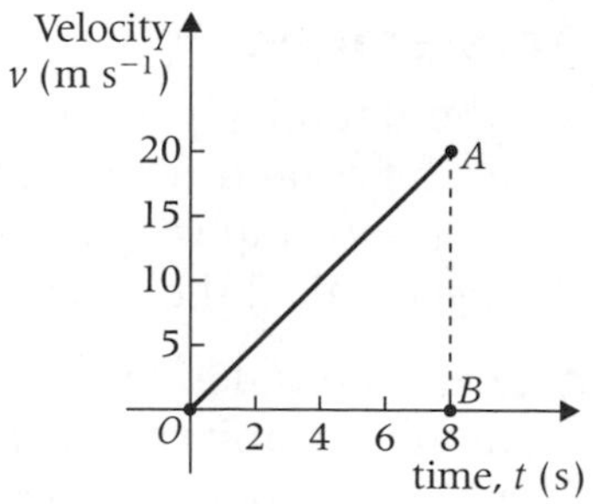

The area under this graph is the area of the triangle OAB, which can be calculated as

$$\begin{aligned}\text{Area} &= \frac{20 \times 8}{2} \\ &= 80\end{aligned}$$

We can compare this with the average speed of the lorry over the 8-second period which would be $10\ \text{m s}^{-1}$. A lorry travelling at $10\ \text{m s}^{-1}$ for 8 seconds would cover 80 metres. Note that this distance is the same as our area.

(This example is not intended to be a "proof", but merely an indication of how the method works.)

> The area under a velocity–time graph represents the distance travelled.

Care must be taken with problems that include both positive and negative velocities. Consider the graph shown here. The shaded area marked A_1 represents a distance travelled in the positive direction, while the area marked A_2 represents a distance travelled in the negative direction. Comparing the sizes of the shaded areas indicates that a greater distance has been travelled in the positive direction than in the negative direction, so that the final displacement will be positive.

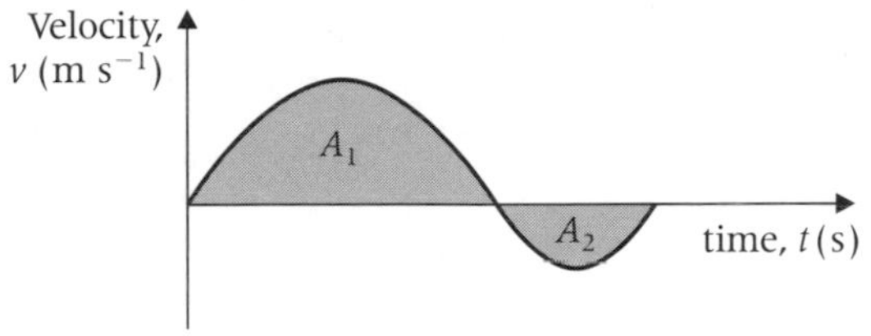

Worked example 1

Alongside a railway track there are marker posts, spaced at kilometre intervals. A train, travelling at constant velocity, is timed to take 2 minutes to travel from one post to the next. After passing the second post, the train slows down uniformly to stop in 0.4 km. Find

(a) the constant velocity of the train

(b) the time it takes to stop

(c) the acceleration of the train.

Solution

(There is a mixture of units in this question so we will use *seconds* for the time and *metres* for distance.)

First we make a sketch of the v–t graph, as shown. Note that BC is a straight line as the train slows down uniformly.

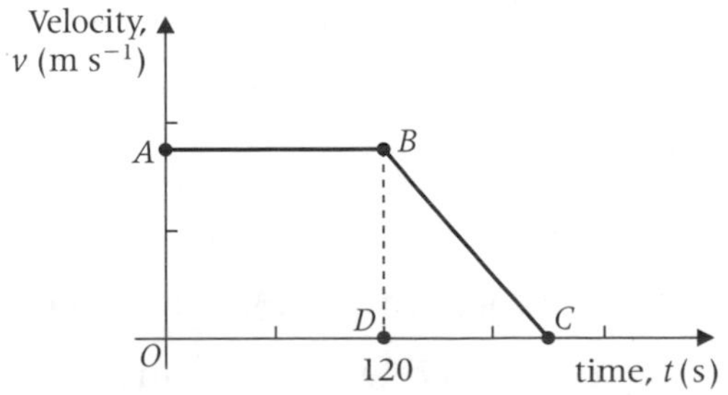

(a) The train travels 1000 m in 120 seconds.
The (constant) speed of the train

$$= \frac{1000}{120} = \frac{25}{3} = 8.33\ \text{m s}^{-1}$$

(to three significant figures)

(b) During the retardation, the train travels 400 m so the area of the triangle BDC must be 400 units, hence

$$\frac{BD \times DC}{2} = 400$$

$$DC = 96$$

Thus the train slows down for 96 seconds.

(c) The acceleration of the train is the gradient of BC.

$$a = \frac{-8.33}{96}$$
$$= -0.0868 \text{ m s}^{-2} \text{ (to three significant figures)}$$

Worked example 2

A cyclist rides along a straight road from X to Y. He starts from rest at X and accelerates uniformly to reach a speed of 10 m s^{-1} in 8 seconds. He travels at this speed for 20 seconds and then decelerates uniformly to stop at Y. If the whole journey takes 40 seconds, sketch a velocity–time graph for the journey.

Use the graph to find:

(a) the initial acceleration

(b) the acceleration on the final stage

(c) the total distance travelled.

Solution

The v–t graph is shown below.

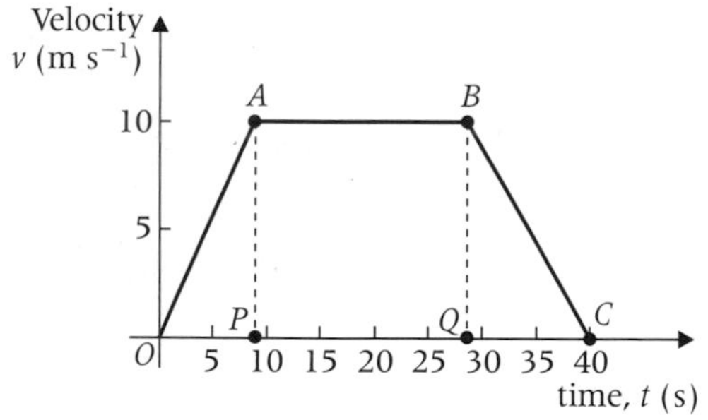

(a) Acceleration $= \frac{10}{8} = 1.25 \text{ m s}^{-2}$

(b) Acceleration $= \frac{-10}{12}$

$$= -\frac{5}{6} \text{ m s}^{-2}$$

We could say that the acceleration is $-\frac{5}{6}$ m s^{-2} or that the cyclist decelerates at $\frac{5}{6}$ m s^{-2}.

(c) Total distance = Area APO + Area $ABQP$ + Area BQC

$$= \frac{8 \times 10}{2} + 20 \times 10 + \frac{12 \times 10}{2}$$

$$= 40 + 200 + 60 = 300\,\text{m}.$$

EXERCISE 2A

1 The graph shows how the velocity of a car changes during a short journey. Find the distance travelled by the car and the acceleration on each stage of its journey.

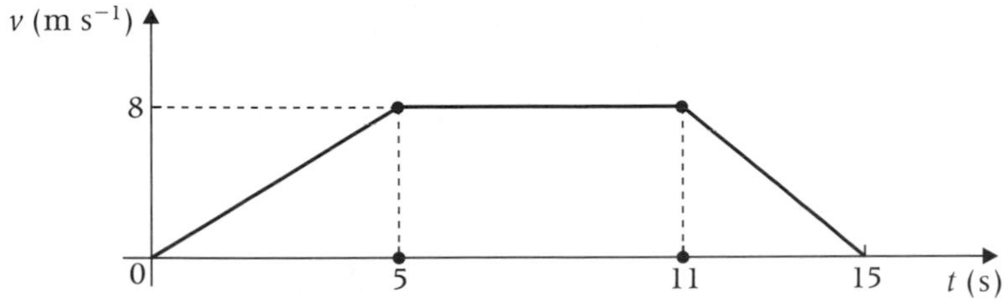

2 This graph shows how the velocity of a cyclist changes over a short period of time. Find the total distance travelled by the cyclist.

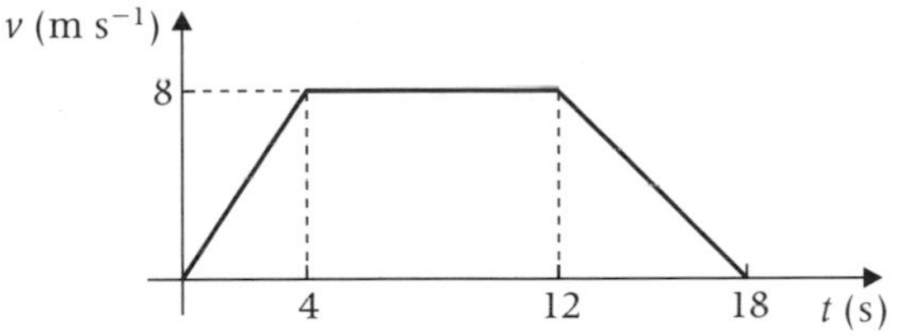

3 Discuss the motion represented by each of the displacement–time graphs shown here.

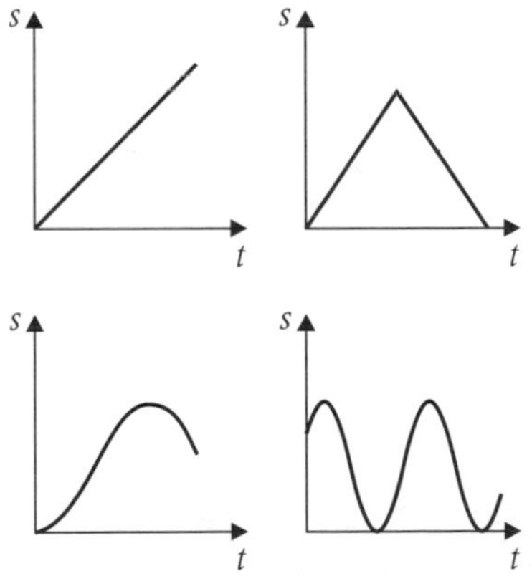

4 Sketch displacement–time and velocity–time graphs for the following.

(a) A car that starts from rest and increases its velocity steadily to $10\,\text{m s}^{-1}$ in 5 seconds. The car holds this velocity for another 10 seconds and then slows steadily to rest in a further 10 seconds.

(b) A ball that is dropped on to a horizontal floor from a height of 3 m. The ball bounces several times before coming to rest.

(c) A person who jumps out of a balloon and falls until the parachute opens. The person then glides steadily to the ground.

5

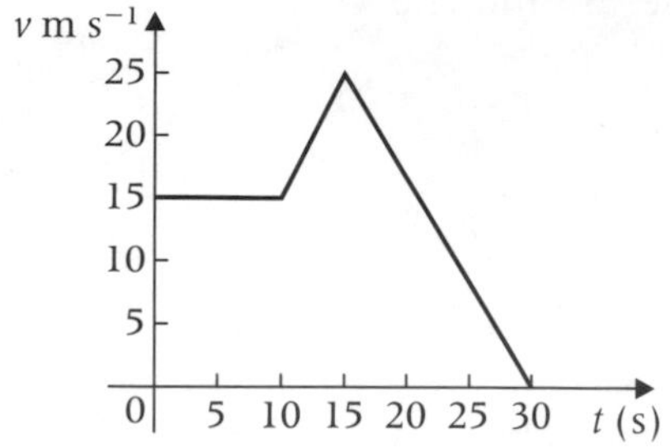

This velocity–time graph illustrates the motion of an object. Calculate the acceleration for each of the following intervals:

(a) $0 < t < 10$

(b) $10 < t < 15$

(c) $15 < t < 30$

Calculate the displacement of the object over the 30 seconds.

6 A car accelerates at 2 m s^{-2} from rest until it reaches a speed of 16 m s^{-1}. It then travels at this speed for 30 seconds, before slowing down and stopping in a further 5 seconds.

Find the total distance travelled by the car. [A]

7

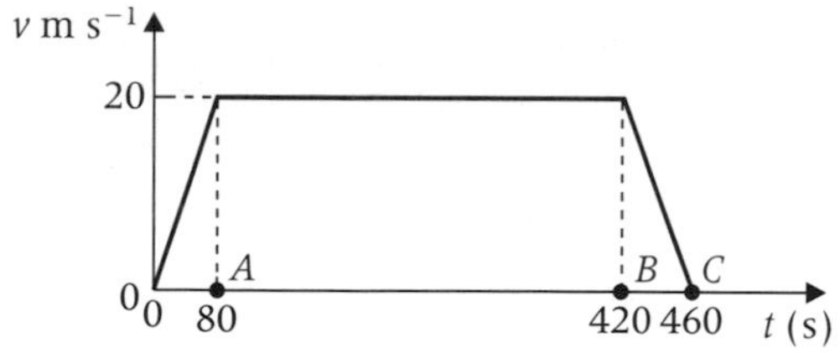

The diagram shows the velocity–time graph for a train which travels from rest in one station to rest at the next station. For each of the time intervals *OA*, *AB* and *BC*, state the value of the train's acceleration.

Calculate the distance between the stations. [A]

8 A train is travelling at a constant speed of 40 m s^{-1}, when the driver sees a warning light. Over the next 1000 m the speed of the train drops to 20 m s^{-1}. The train travels at this speed for

5 minutes. The speed returns to 40 m s^{-1} after a further 5 minutes. Assume that the acceleration of the train is constant on each stage of its journey.

(a) Find the total distance travelled by the train, while its speed is less than its normal operating speed of 40 m s^{-1}.

(b) The train **would normally** have travelled this distance at a constant 40 m s^{-1}. Find the time by which it was delayed. [A]

9 A tram travelling along a straight track starts from rest and accelerates uniformly for 15 seconds. During this time it travels 135 metres. The tram now maintains a constant speed for a further 1 minute. It is finally brought to rest decelerating uniformly over a distance of 90 metres. Calculate the tram's acceleration and deceleration during the first and last stages of the journey. Also find the time taken and the distance travelled for the whole journey. [A]

10 A train travelling at 50 m s^{-1} applies its brakes on passing a yellow signal at a point A and decelerates uniformly, with a deceleration of 1 m s^{-2}, until it reaches a speed of 10 m s^{-1}. The train then travels for 2 km at the uniform speed of 10 m s^{-1} before passing a green signal. On passing the green signal the train accelerates uniformly, with acceleration 0.2 m s^{-2}, until it finally reaches a speed of 50 m s^{-1} at a point B. Find the distance AB and the time taken to travel that distance. [A]

11 Two sprinters compete in a 100 m race, crossing the finishing line together after 12 seconds. The two models, A and B, as described below, are models for the motions of the two sprinters.

Model A. The sprinter accelerates from rest at a constant rate for 4 seconds and then travels at a constant speed for the rest of the race.

Model B. The sprinter accelerates from rest at a constant rate until reaching a speed of 9 m s^{-1} and then travels at this speed for the rest of the race.

(a) For model A, find the maximum speed and the initial acceleration of the sprinter.

(b) For model B, find the time taken to reach the maximum speed and the initial acceleration of the sprinter.

(c) Sketch a distance–time graph for each of the two sprinters on the same set of axes. Describe how the distance between the two sprinters varies through the race. [A]

12 A car travels a total distance of 430 m in a time of 25 seconds. During this time, the car accelerates from rest at 5 m s^{-2} for 4 seconds, then travels at a constant speed and finally slows down, with constant deceleration, until it stops.

(a) Find the distance travelled by the car in the first 4 seconds and the speed of the car at the end of this time.

(b) Find the time for which the car travels at a constant speed and the deceleration during the final stage of the car's motion. [A]

2.3 Motion under constant acceleration

There are several simple formulae which can be used to solve problems that involve motion under **constant** (or **uniform**) acceleration.

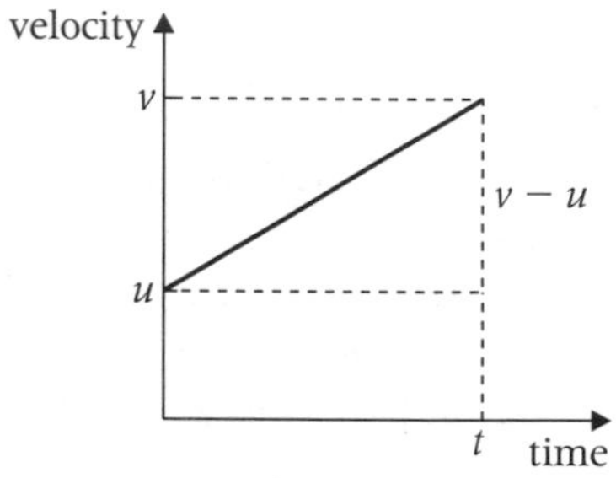

The diagram is a velocity–time graph for the motion of an object with initial velocity u and final velocity v after t seconds has elapsed.

The gradient of the line is equal to the acceleration and is calculated from the expression

$$\frac{v-u}{t}$$

Hence

$$a = \frac{v-u}{t}$$

which can be rewritten as

$$v = u + at.$$

The area under the velocity–time graph is equal to the displacement of the object. Using the rule for the area of a trapezium gives

$$s = \frac{1}{2}(u+v)t$$

Worked example 3

A motorbike accelerates at a constant rate of 3 m s^{-2}. Calculate

(a) the time taken to accelerate from 18 kph to 45 kph

(b) the distance, in metres, covered in this time.

Solution

We can use the equation $v = u + at$ to find the time and then the equation $s = \frac{1}{2}(u + v)t$ to find the distance travelled. But first the units for speed must be converted to m s^{-1}.

$$18 \text{ kph} = \frac{18 \times 1000}{3600} = 5 \text{ m s}^{-1}$$

and similarly

$$45 \text{ kph} = 12.5 \text{ m s}^{-1}$$

(a) Using $v = u + at$, with $u = 5$, $v = 12.5$ and $a = 3$ gives

$$v = u + at$$

$$12.5 = 5 + 3t$$

$$t = \frac{7.5}{3} = 2.5 \text{ seconds}$$

(b) Using $s = \frac{1}{2}(u + v)t$ with $u = 5$, $v = 12.5$ and $t = 2.5$ gives

$$s = \frac{1}{2}(u + v)t$$

$$= \frac{1}{2}(12.5 + 5) \times 2.5$$

$$= 21.875 \text{ m}$$

Two more useful formulae

We can write the equation $v = u + at$ in the form

$$t = \frac{v - u}{a}$$

and substitute it into $s = \frac{1}{2}(u + v)t$, then

$$s = \frac{1}{2}(u + v)\left(\frac{v - u}{a}\right)$$

$$= \frac{v^2 - u^2}{2a}$$

which we can rearrange to give

$$v^2 = u^2 + 2as$$

Similarly, using $v = u + at$ to substitute for v in the equation

$$s = \frac{1}{2}(u + v)t$$

gives

$$s = \frac{(u + u + at)t}{2}$$

or

$$s = ut + \tfrac{1}{2}at^2$$

This gives four different constant acceleration formulae. The appropriate one can be selected to fit the data available in the problem that you have to solve.

> When using these formulae it is important to remember that they only apply to cases where the acceleration is constant or can be assumed to be constant.

Worked example 4

A car accelerates from a velocity of $16\ \text{m s}^{-1}$ to a velocity of $40\ \text{m s}^{-1}$ in a distance of 500 m. Find the acceleration of the car.

Solution

Using the equation $v^2 = u^2 + 2as$, with $u = 16$, $v = 40$ and $s = 500$ gives

$$40^2 = 16^2 + 2 \times a \times 500$$

$$a = \frac{1600 - 256}{1000} = 1.344\ \text{m s}^{-2}$$

Worked example 5

A car decelerates from a velocity of $36\ \text{m s}^{-1}$. The magnitude of the deceleration is $3\ \text{m s}^{-2}$. Calculate the time required to travel a distance of 162 m.

Solution

When an object is slowing down (decelerating) we can use the constant acceleration equations, but with a negative value for a. In this case $a = -3$.

We require the time, t seconds to travel 162 m.

Using the constant acceleration equation $s = ut + \frac{1}{2}at^2$, with $s = 162$, $u = 36$ and $a = -3$ gives

$$162 = 36t + \tfrac{1}{2}(-3)t^2$$

Rearranging this gives this quadratic equation:

$$1.5t^2 - 36t + 162 = 0$$

Dividing by 1.5 gives

$$t^2 - 24t + 108 = 0$$

And factorising gives

$$(t - 6)(t - 18) = 0$$

so that

$$t = 6 \text{ or } t = 18$$

The first answer, of 6 seconds, is the required one.

The second answer would give the time that the displacement was 162 m for the second time. In this case the car would be moving in the opposite direction.

Worked example 6

A cyclist is initially travelling at $10\ \text{m s}^{-1}$, when she applies her brakes. Assume that her acceleration remains constant at $-0.8\ \text{m s}^{-2}$ until she stops. Find the distance that she travels before stopping and the time that it takes her to stop.

Solution

To find the distance that she travels use the formula $v^2 = u^2 + 2as$, with $v = 0$, $u = 10$ and $a = -0.8$, which gives

$$v^2 = u^2 + 2as$$

$$0^2 = 10^2 + 2 \times (-0.8)s$$

$$0 = 100 - 1.6s$$

$$s = \frac{100}{1.6} = 62.5\ \text{m}$$

Now the time taken to stop can be found using the formula $s = \frac{1}{2}(u + v)t$, with $s = 62.5$, $u = 10$ and $v = 0$, to give

$$s = \frac{1}{2}(u + v)t$$

$$62.5 = \frac{1}{2}(10 + 0)t$$

$$62.5 = 5t$$

$$t = 12.5 \text{ seconds}$$

EXERCISE 2B

1 A car accelerates at $2\ \text{m s}^{-2}$ from rest for 10 seconds.

(a) Find the distance travelled by the car and the speed it reaches.

(b) After the 10 seconds its acceleration changes to $0.5\ \text{m s}^{-2}$ and then remains constant for a further 5 seconds. Find the speed of the car and the total distance that it has travelled at the end of the 15 seconds.

2 A car accelerates from $10\ \text{m s}^{-1}$ to $20\ \text{m s}^{-1}$ as it travels 500 m.

(a) Find the acceleration of the car.

(b) Find the time taken by the car to travel the 500 m.

3 A car accelerates uniformly from 5 m s^{-1} to 12 m s^{-1}, in a 10-second period of time.

(a) Find the acceleration of the car.

(b) Find the distance travelled by the car.

4 A lift rises from rest, accelerating at a constant rate until it reaches a speed of 1.6 m s^{-1} after 8 seconds.

(a) Find the acceleration of the lift.

(b) The lift continues to accelerate for a further 2 seconds. Find the distance that the lift has now risen.

(c) The lift then shows down, at a constant rate, and stops after a further 5 seconds. Find the total distance travelled by the lift.

5 A car accelerates uniformly from a speed of 50 kph to a speed of 80 kph in 20 seconds.

Calculate the acceleration in m s^{-2}.

6 For the car in Question 5, calculate the distance travelled during the 20 seconds.

7 A van travelling at 40 mph skids to a halt in a distance of 15 m. Find the acceleration of the van and the time taken to stop, assuming that the deceleration is uniform. (Assume 1 mile = 1600 m.)

8 A train signal is placed so that a train can decelerate uniformly from a speed of 96 kph to come to rest at the end of a platform. For passenger comfort the deceleration must be no greater than 0.4 m s^{-2}. Calculate

(a) the shortest distance the signal can be from the platform

(b) the shortest time for the train to decelerate.

9 A rocket is travelling with a velocity of 80 m s^{-1}. The engines are switched on for 6 seconds and the rocket accelerates uniformly at 40 m s^{-2}. Calculate the distance travelled over the 6 seconds.

10 The world record for the men's 60 m race was 6.41 seconds.

(a) Assuming that the race was carried out under constant acceleration, calculate the acceleration of the runner and his speed at the end of the race.

(b) Now assume that in a 100 m race the runner accelerates for the first 60 m and completes the race by running the next 40 m at the speed you calculated in **(a)**.

Calculate the time for the athlete to complete the race.

11 The world record for the men's 100 m was 9.83 seconds. Assume that the last 40 m was run at constant speed and that the acceleration during the first 60 m was constant.

(a) Calculate this speed.

(b) Calculate the acceleration of the athlete.

2

12 Telegraph poles, 40 m apart stand alongside a railway line. The times taken for a locomotive to pass the two gaps between three consecutive poles are 2.5 seconds and 2.3 seconds, respectively. Calculate the acceleration of the train and the speed past the first post.

13 A set of traffic lights covers road repairs on one side of a road in a 30 mph speed limit area. The traffic lights are 80 m apart so time must be allowed to delay the light changing from green to red. Assuming that a car accelerates at 2 m s^{-2} what is the least this time delay should be?

14 A train starts from rest and moves with constant acceleration $\frac{1}{3}\text{ m s}^{-2}$ for 2 minutes. For the next 4 minutes the train moves with zero acceleration, after which a uniform retardation of 2 m s^{-2} brings it to rest. Find the total distance travelled by the train from starting to stopping. [A]

15 As a train leaves a station it accelerates, from rest, at 0.8 m s^{-2} for 30 seconds, travels at a constant speed for the next 5 minutes and then slows down, stopping in 20 seconds at a second station.

(a) Find the maximum speed of the train.

(b) Find the distance travelled by the train between the stations, clearly stating any assumptions that you have made. [A]

16 As a lift moves upwards from rest it accelerates at 0.8 m s^{-2} for 2 seconds, then travels 4 m at constant speed and finally slows down, with a constant deceleration, stopping in 3 seconds.

Find the total distance travelled by the lift and the total time taken. [A]

17 Two cars A and B are initially at rest side by side. A starts off on a straight track with an acceleration of 2 m s^{-2}. Five seconds later B starts off on a parallel track to A, with acceleration 3.125 m s^{-2}.

(a) Calculate the distance travelled by A after 5 seconds.

(b) Calculate the time taken for B to catch up A.

(c) Find the speeds of A and B at that time.

18 Two cars are initially 36 m apart travelling in the same direction along a straight, horizontal road. The car in front is initially travelling at $10\ \text{m s}^{-1}$, but decelerating at $2\ \text{m s}^{-2}$. The other car travels at a constant $15\ \text{m s}^{-1}$.

(a) Model the cars as particles. By finding the distance travelled by each car after t seconds, show that the distance between the two cars is $36 - 5t - t^2$ metres. Find when they would collide if neither car takes avoiding action.

(b) Would it be necessary to revise your answers to part **(a)** if the cars were not modelled as particles? Give reasons to support your answer. [A]

19 Two humps are to be installed on a road to prevent traffic reaching speeds of greater than $12\ \text{m s}^{-1}$ between the humps. Assume that:

I the speed of cars when they cross the humps is effectively zero

II after crossing a hump they accelerate at $3\ \text{m s}^{-2}$ until they reach a speed of $12\ \text{m s}^{-1}$

III as soon as they reach a speed of $12\ \text{m s}^{-1}$ they decelerate at $6\ \text{m s}^{-2}$ until they stop.

(a) A simple model ignores the lengths of the cars. Use this to find the distance between the humps.

(b) One factor that has not been taken into account is the length of the cars. Revise your answer to part **(a)** to take this into account, giving your answer to the nearest metre. You must state clearly any assumptions that you make. [A]

2.4 Motion under gravity

For many centuries it was believed that:

(a) heavier bodies fell faster than light ones, and

(b) the speed of a falling body was constant all through its motion.

Galileo Galilei (1564–1642) was the first person to state clearly (and to demonstrate) that all objects fall with the same acceleration.

Modern scientific instruments determine the acceleration of falling bodies as values in the region of $9.81\ \text{m s}^{-2}$, although the value varies slightly at different places on the earth's surface, and at different altitudes. The symbol used to represent this "acceleration due to gravity" is g.

The value of g is sometimes approximated to $10\,\text{m s}^{-2}$, but in this book we shall normally use $9.8\,\text{m s}^{-2}$ unless stated otherwise. Since this acceleration acts towards the earth's surface, its sign must always be opposite to that of any velocities that are upwards.

When solving problems involving motion under gravity (ignoring any air resistance at this stage) the formulae for motion with constant acceleration may be used.

Worked example 7

A ball is projected vertically upwards with an initial speed of $30\,\text{m s}^{-1}$. Calculate the maximum height reached.

Solution

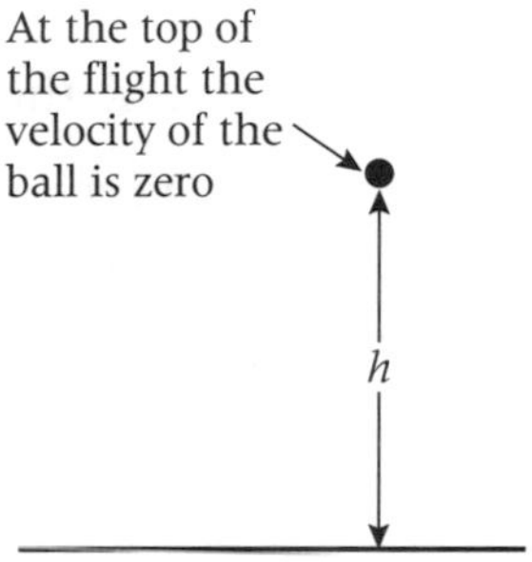
At the top of the flight the velocity of the ball is zero
h

At the top of the ball's flight, its speed will be zero.

Take the upwards direction as positive, so the acceleration will be

$$a = -g = -9.8\,\text{m s}^{-2}.$$

Using $v^2 = u^2 + 2as$ gives

$$0 = 30^2 - 2 \times 9.8 \times h$$

where h is the maximum height reached.

Thus

$$h = \frac{900}{19.6} = 45.9\,\text{m (to three significant figures).}$$

Worked example 8

A stone is fired vertically upwards with an initial speed of $10\,\text{m s}^{-1}$ from a catapult. Calculate the interval between the two times when the ball is 5 m above the point of release.

Solution

The stone is at a height of 5 m on its upward journey and again on the way down. The neatest way to solve the problem is as follows.

Using $s = ut + \frac{1}{2}at^2$ with

$$a = -g = -9.8\,\text{m s}^{-2}$$

$$5 = 10t - 0.5 \times 9.8 \times t^2$$

where t seconds is the time passed since the stone was thrown up.

$$4.9t^2 - 10t + 5 = 0$$

Using the quadratic equation formula

$$t = \frac{10 \pm \sqrt{2}}{9.8} = 0.876 \text{ or } 1.165$$

the required time interval is $1.165 - 0.876 = 0.289$ seconds.

EXERCISE 2C

1 A ball is dropped from rest at a height of 2 m. Find the time that the ball takes to fall to the ground, if it is:

(a) on earth

(b) on the moon, where $g = 1.6\ \text{m s}^{-2}$.

2 A ball, that is initially at rest, falls from a height of 3 m to the ground.

(a) Find the time that the ball takes to fall this distance.

(b) Find the speed of the ball when it hits the ground.

3 A ball is thrown upwards with an initial speed of $14.7\ \text{m s}^{-1}$ from a height of 1 m.

(a) Find the time that it takes the ball to reach its maximum height.

(b) Find the maximum height of the ball.

(c) Find the speed of the ball when it hits the ground.

4 A rocket rises from ground level to a height of 100 m in 10 seconds. Assume that the acceleration of the rocket is constant and that it starts at rest.

(a) Find the acceleration of the rocket and its speed at a height of 100 m.

After these 10 seconds the rocket runs out of fuel, but continues to move vertically under the influence of gravity.

(b) Find the maximum height of the rocket.

5 The diagram shows three positions of a ball which has been thrown upwards with a velocity of $u\ \text{m s}^{-1}$

Position A is the initial position.

Position B is halfway up.

Position C is at the top of the motion.

Copy the diagram and for each position put on arrows where appropriate to show the direction of the velocity.

On the same diagram put on arrows to show the direction of the acceleration.

6 A ball is dropped on to level ground from a height of 20 m.

(a) Calculate the time taken to reach the ground.

The ball rebounds with half the speed it strikes the ground.

(b) Calculate the time taken to reach the ground a second time.

7 A stone is thrown down from a high building with an initial velocity of 4 m s^{-1}. Calculate the time required for the stone to drop 30 m and its velocity at this time.

8 A ball is thrown vertically upwards from the top of a cliff which is 50 m high. The initial velocity of the ball is 25 m s^{-1}. Calculate the time taken to reach the bottom of the cliff and the velocity of the ball at that instant.

9 One stone is thrown upwards with a speed of 2 m s^{-1} and another is thrown downwards with a speed of 2 m s^{-1}. Both are thrown at the same time from a window 5 m above ground level.

(a) Which hits the ground first?

(b) Which is travelling fastest when it hits the ground?

(c) What is the total distance travelled by each stone?

10 A ball is thrown vertically upwards with an initial velocity of 30 m s^{-1}. One second later, another ball is thrown upwards with an initial velocity of $u\text{ m s}^{-1}$. The particles collide after a further 2 seconds. Find the value of u.

11 When a ball hits the ground it rebounds with half of the speed that it had when it hit the ground. If the ball is dropped from a height h, calculate the height to which it rebounds.

12 A small canister is attached to a helium-filled balloon and released from rest at ground level. After 4 seconds it is moving vertically upwards at 6 m s^{-1}.

(a) Find the height of the balloon and canister after 4 seconds, stating clearly any assumptions that you make.

When the balloon reaches a height of 27 m it bursts.

(b) Find the maximum height reached by the canister. [A]

13 A tennis ball is hit so that it moves vertically downwards from a height of 1 m with an initial speed of 5 m s^{-1}. When it hits the ground it rebounds vertically with half the speed it had when it hit the ground.

(a) Find the height to which it rebounds.

(b) State whether this is likely to be an under-estimate or an over-estimate, giving reasons to support your answer. [A]

Key point summary

Formulae to learn

$v = u + at$

$s = \frac{1}{2}(u + v)t$

$v^2 = u^2 + 2as$

$s = ut + \frac{1}{2}at^2$

1 The gradient of a displacement–time graph gives the velocity. pp8–9

2 The gradient of a velocity–time graph gives the acceleration. p10

3 The area under a velocity–time graph can be used to find the displacement. p11

4 Only use the constant acceleration formulae when the acceleration is constant or can be assumed to be constant. p18

5 All objects accelerate in the same way when falling under the influence of gravity alone. p22

6 The acceleration due to gravity is $9.8\ \text{m s}^{-2}$. p22

Test yourself

What to review

If your answer is incorrect – review

1 A car accelerates uniformly from rest to $20\ \text{m s}^{-1}$ in 25 seconds. It travels at this speed for 1.5 minutes and then slows down, stopping after a further 35 seconds. — *Section 2.2*

(a) Draw a velocity–time graph and use it to find the total distance travelled by the car.

(b) Calculate the acceleration of the car on each stage of its journey.

---

2 The velocity of a car increases from $5\ \text{m s}^{-1}$ to $25\ \text{m s}^{-1}$ as it travels a distance of 100 m. Assume that the acceleration of the car is constant. — *Section 2.3*

(a) Find the acceleration of the car.

(b) Find the speed of the car when it has travelled 50 m.

(c) Find the time it takes for the car to travel the 100 m.

---

3 A stone is thrown upwards from a height of 2 m above ground level. It reaches a maximum height of 5 m above ground level. — *Section 2.4*

(a) Find the initial velocity of the stone.

(b) Find the velocity of the stone when it hits the ground.

(c) How long is the stone in the air?

Test yourself ANSWERS

1 **(a)** 2400 m, **(b)** $0.8\,\mathrm{m\,s^{-2}}$, $0\,\mathrm{m\,s^{-2}}$, $-\frac{4}{7}\,\mathrm{m\,s^{-2}}$.

2 **(a)** $3\,\mathrm{m\,s^{-2}}$, **(b)** $18.0\,\mathrm{m\,s^{-1}}$, **(c)** $6\frac{2}{3}$ s.

3 **(a)** $7.67\,\mathrm{m\,s^{-1}}$, **(b)** $9.90\,\mathrm{m\,s^{-1}}$, **(c)** 1.79 s.

CHAPTER 3

Kinematics in two and three dimensions

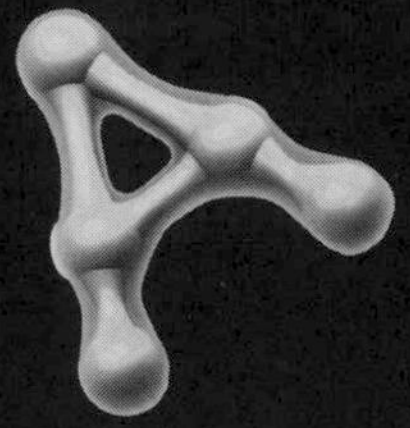

Learning objectives

After studying this chapter you should be able to:

- plot and interpret paths given a position vector
- write positions, velocities and accelerations in the form $x\mathbf{i} + y\mathbf{j} + z\mathbf{k}$ or $x\mathbf{i} + y\mathbf{j}$
- find magnitudes and directions of vectors
- apply and use the constant acceleration equations in two or three dimensions.

3.1 Introduction

In this chapter the idea of using constant acceleration is extended into two and three dimensions. Before doing this, we need to be able to describe motion in more than one dimension. This is done using vectors, which we will consider first in this chapter.

3.2 Describing motion in two and three dimensions

We will first consider two dimensions and then extend the ideas to three dimensions.

The first key step is to define an origin or reference point. The position of an object is then described relative to this point. Often the origin will be the initial position of an object.

Secondly we define two perpendicular unit vectors. These have length and are directed at right angles to each other.

The diagram at the top of the next page shows an origin O and two unit vectors $\mathbf{i}$ and $\mathbf{j}$.

In addition the diagram shows some points. We can describe the position of each point by using the unit vectors $\mathbf{i}$ and $\mathbf{j}$. For example the position of the point A can be written as:

$$\mathbf{r}_A = 4\mathbf{i} + 3\mathbf{j}$$

Similarly

$$\mathbf{r}_B = 5\mathbf{i} - 2\mathbf{j}$$
$$\mathbf{r}_C = -5\mathbf{i} - 3\mathbf{j}$$
$$\mathbf{r}_D = -3\mathbf{i} + 2\mathbf{j}$$

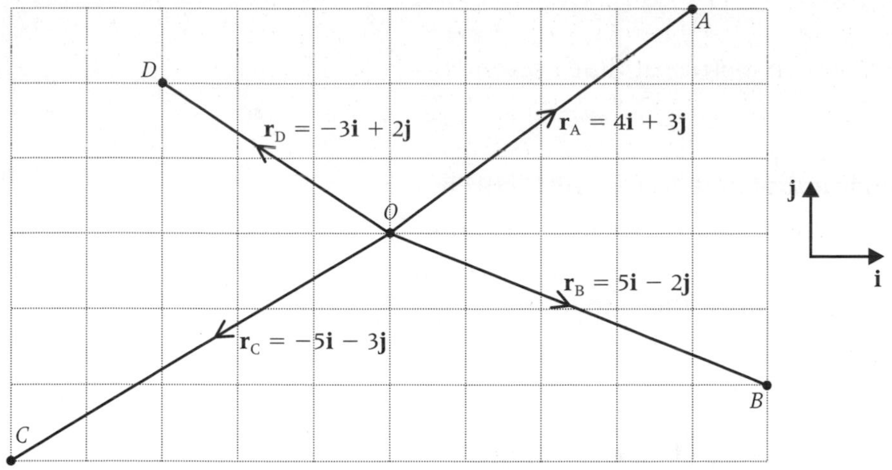

3

It is possible to describe the position of a particle in terms of the time, t, that it has been moving. For example we could write:

$$\mathbf{r} = (4t + 3)\mathbf{i} + (t^2 - 4t)\mathbf{j}$$

Then the position vector $\mathbf{r}$ can be found for any value of t and the path of the object can be drawn or described.

For two-dimensional motion we usually write:

$$\mathbf{r} = x(t)\mathbf{i} + y(t)\mathbf{j}$$

where $\mathbf{i}$ and $\mathbf{j}$ lie in the plane that the motion takes place in. If this is a vertical plane it is the normal convention that $\mathbf{i}$ is horizontal and $\mathbf{j}$ is vertical. The diagram shows the position of a particle.

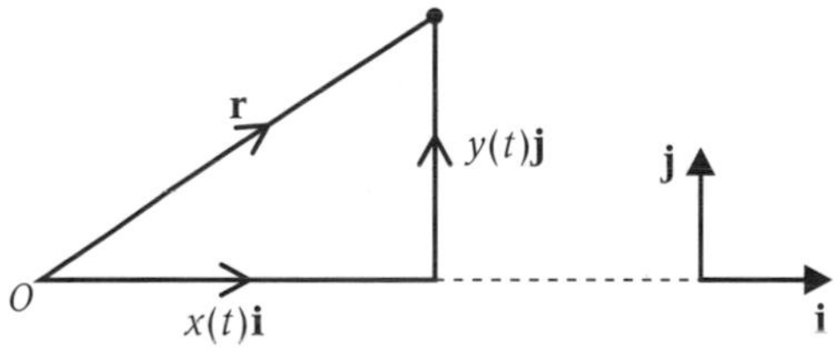

$$\mathbf{r} = (4t + 3)\mathbf{i} + (t^2 - 4t)\mathbf{j}$$

Then the position vector $\mathbf{r}$ can be found for any value of t and the path of the object can be drawn or described.

For three-dimensional motion we usually write:

$$\mathbf{r} = x(t)\mathbf{i} + y(t)\mathbf{j} + z(t)\mathbf{k}$$

where $\mathbf{i}$ and $\mathbf{j}$ lie in a horizontal plane and $\mathbf{k}$ is a vertical unit vector. Note that these three unit vectors are all perpendicular as shown in the diagram.

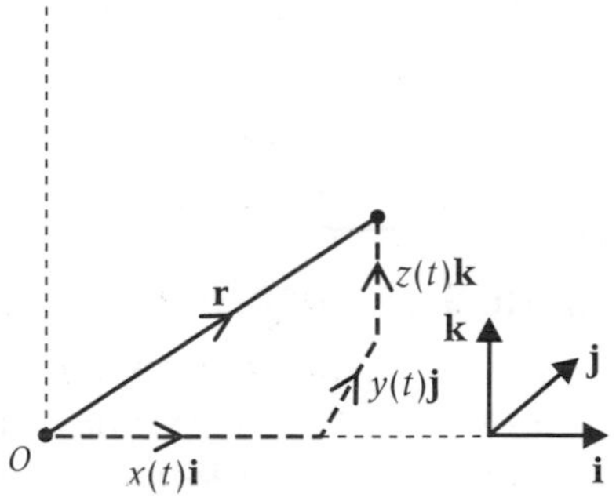

Note that $\mathbf{r}$ is referred to as the position vector of the object that is moving.

Worked example 1

A ball is thrown so that its position, in metres, at time t seconds is given by

$$\mathbf{r} = 6t\mathbf{i} + (1 + 8t - 5t^2)\mathbf{j}$$

where **i** and **j** are horizontal and vertical unit vectors, respectively.

(a) Find the position of the particle when $t = 0, 0.5, 1, 1.5$ and 2 seconds.

(b) Plot the positions in part **(a)** and draw the path of the particle.

Solution

(a) The table below shows how the values of t are substituted.

t	$\mathbf{r}$
0	$6 \times 0\mathbf{i} + (1 + 8 \times 0 - 5 \times 0^2)\mathbf{j} = 0\mathbf{i} + 1\mathbf{j}$
0.5	$6 \times 0.5\mathbf{i} + (1 + 8 \times 0.5 - 5 \times 0.5^2)\mathbf{j} = 3\mathbf{i} + 3.75\mathbf{j}$
1	$6 \times 1\mathbf{i} + (1 + 8 \times 1 - 5 \times 1^2)\mathbf{j} = 6\mathbf{i} + 4\mathbf{j}$
1.5	$6 \times 1.5\mathbf{i} + (1 + 8 \times 1.5 - 5 \times 1.5^2)\mathbf{j} = 9\mathbf{i} + 1.75\mathbf{j}$
2	$6 \times 2\mathbf{i} + (1 + 8 \times 2 - 5 \times 2^2)\mathbf{j} = 12\mathbf{i} - 3\mathbf{j}$

(b) These positions can then be plotted and a curve drawn to show the path of the ball.

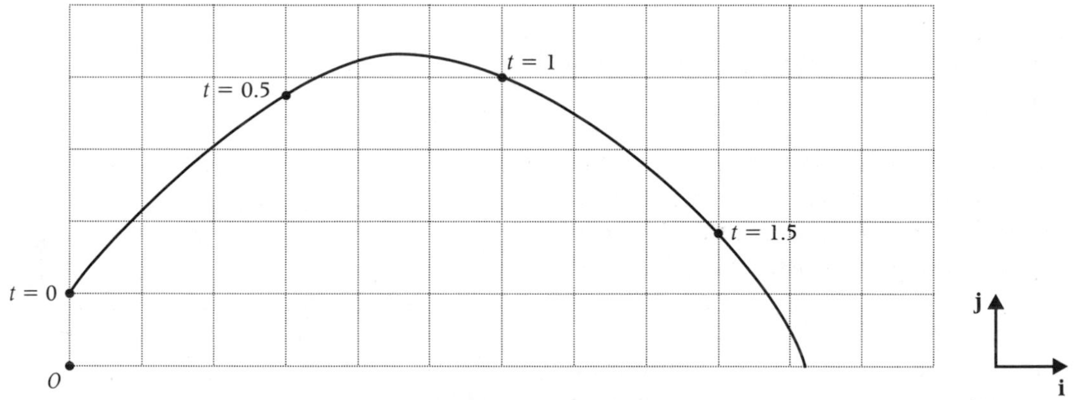

If the origin is at ground level we can note, from the diagram, that:

(i) the ball is thrown from a height of 1 m

(ii) the ball reaches a maximum height of 4.2 m

(iii) the horizontal distance travelled by the ball until it hits the ground is 10.3 m.

Worked example 2

Two boats move on a pond. They both start at the same place. One follows a curved path and the other travels along a straight line. The positions, in metres, of the boats, at time t seconds are given by:

$$\mathbf{r}_A = 2t\mathbf{i} + 4t\mathbf{j}$$

$$\mathbf{r}_B = (8t - t^2)\mathbf{i} + 4t\mathbf{j}$$

where **i** and **j** are unit vectors directed east and north, respectively.

(a) Find the positions of the boats when $t = 0, 2, 4$ and 6 seconds.

(b) What happens when $t = 6$?

(c) Plot the paths of the two boats.

Solution

(a) The table below shows how the positions are calculated, by substituting the required values of t.

t	$\mathbf{r}_A$	$\mathbf{r}_B$
0	$2 \times 0\mathbf{i} + 4 \times 0\mathbf{j} = 0\mathbf{i} + 0\mathbf{j}$	$(8 \times 0 - 0^2)\mathbf{i} + 4 \times 0\mathbf{j} = 0\mathbf{i} + 0\mathbf{j}$
2	$2 \times 2\mathbf{i} + 4 \times 2\mathbf{j} = 4\mathbf{i} + 8\mathbf{j}$	$(8 \times 2 - 2^2)\mathbf{i} + 4 \times 2\mathbf{j} = 12\mathbf{i} + 8\mathbf{j}$
4	$2 \times 4\mathbf{i} + 4 \times 4\mathbf{j} = 8\mathbf{i} + 16\mathbf{j}$	$(8 \times 4 - 4^2)\mathbf{i} + 4 \times 4\mathbf{j} = 16\mathbf{i} + 16\mathbf{j}$
6	$2 \times 6\mathbf{i} + 4 \times 6\mathbf{j} = 12\mathbf{i} + 24\mathbf{j}$	$(8 \times 6 - 6^2)\mathbf{i} + 4 \times 6\mathbf{j} = 12\mathbf{i} + 24\mathbf{j}$

(b) When $t = 6$ the boats both have the same position at the same time and so will collide.

(c) The paths of the boats are shown below.

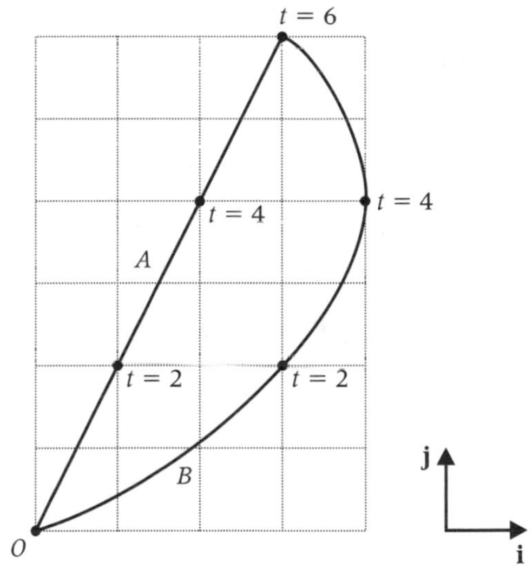

Worked example 3

A boat moves so that its position, in metres, at time t seconds is given by

$$\mathbf{r} = (450 - 5t)\mathbf{i} + (4t + 100)\mathbf{j}$$

where $\mathbf{i}$ and $\mathbf{j}$ are unit vectors that are directed east and north, respectively. A rock has position $250\mathbf{i} + 200\mathbf{j}$.

(a) Calculate the time when the boat is due north of the rock.

(b) Calculate the time when the boat is due east of the rock.

Solution

(a) When the boat is due north of the rock, its position will be $250\mathbf{i} + y\mathbf{j}$, where y is a constant. Equating the $\mathbf{i}$ components of the position vectors gives

$$250 = 450 - 5t$$

$$t = \frac{200}{5} = 40 \text{ seconds}$$

At this time the position of the boat is $250\mathbf{i} + 260\mathbf{j}$.

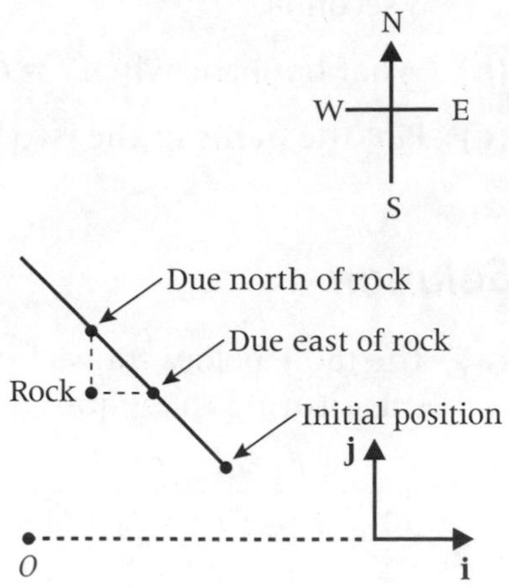

(b) When the boat is due east of the rock, its position will be $x\mathbf{i} + 200\mathbf{j}$, where x is a constant. Equating the $\mathbf{j}$ components of the position vectors gives

$$200 = 100 + 4t$$

$$t = \frac{100}{4} = 25 \text{ seconds}$$

At this time the position of the boat will be $325\mathbf{i} + 200\mathbf{j}$.

EXERCISE 3A

1 A golf ball is hit so that its position, in metres, at time t seconds is given by

$$\mathbf{r} = 30t\mathbf{i} + (25t - 4.9t^2)\mathbf{j}$$

where $\mathbf{i}$ and $\mathbf{j}$ are horizontal and vertical unit vectors, respectively.

(a) Find the position of the ball when $t = 0, 1, 2, 3, 4, 5$ and 6 seconds.

(b) Plot the path of the ball.

(c) From your plot estimate the horizontal distance travelled by the ball when it hits the ground.

2 A ball moves so that its position vector, in metres, at time t seconds is given by:

$$\mathbf{r} = 5t\mathbf{i} + (1 + 4t - 5t^2)\mathbf{j}.$$

where $\mathbf{i}$ and $\mathbf{j}$ are horizontal and vertical unit vectors, respectively.

(a) Find the position of the ball when $t = 0$, $t = 0.2$, $t = 0.4$, $t = 0.6$ and $t = 1$.

(b) Use your answers to **(a)** to sketch the path of the ball.

3 Two children, A and B, run so that their position vectors in metres at time t seconds are given by:

$$\mathbf{r}_A = t\mathbf{i} + t\mathbf{j} \quad \text{and} \quad \mathbf{r}_B = (4 + 4t - t^2)\mathbf{i} + t\mathbf{j}$$

where **i** and **j** are unit vectors directed east and north, respectively.

Plot the paths of the two children for $0 \leq t \leq 4$. What happens when $t = 4$?

4 A boat moves so that its position, in metres, at time t seconds is given by

$$\mathbf{r} = (2t + 6)\mathbf{i} + (3t - 9)\mathbf{j}$$

where **i** and **j** are unit vectors that are directed east and north respectively. A light house has position $186\mathbf{i} + 281\mathbf{j}$.

(a) Find the position of the boat, when $t = 0$, 40, 80 and 120 seconds.

(b) Plot the path of the boat.

(c) From your plot of the path find the shortest distance between the boat and the light house.

(d) Calculate the time when the boat is due south of the lighthouse and the time when the boat is due east of the lighthouse.

5 A bullet is fired from a rifle, so that its position, in metres, at time t seconds is given by

$$\mathbf{r} = 180t\mathbf{i} + (1.225 - 4.9t^2)\mathbf{j}$$

where **i** and **j** are horizontal and vertical unit vectors, respectively, and the origin is at ground level.

(a) Find the initial height of the bullet.

(b) Find the time when the bullet hits the ground.

(c) Find the horizontal distance travelled by the bullet.

6 A boomerang is thrown. As it moves its position, in metres, at time t seconds is modelled by

$$\mathbf{r} = (10t - t^2)\mathbf{i} + 4t\mathbf{j} + (2 + t - t^2)\mathbf{k}$$

where **i** and **j** are perpendicular unit vectors and **k** is a vertical unit vector. The origin is at ground level.

Find the position of the boomerang when it hits the ground.

3.3 Expressing quantities as vectors

Not all quantities such as positions, velocities or accelerations will be expressed in the form $a\mathbf{i} + b\mathbf{j}$. For example a position may be expressed in terms of a distance and a bearing. In this section we see how to express these quantities in this form.

Consider a point that is at a distance d from the origin as shown in the diagram. If the angle between the unit vector **i** and the vector representing the position is θ, then we can write

$$\mathbf{r} = d\cos\theta\,\mathbf{i} + d\sin\theta\,\mathbf{j}$$

If **i** and **j** are horizontal and vertical, respectively, we would say that the horizontal component of the position vector is $d\cos\theta$ and that the vertical component is $d\sin\theta$

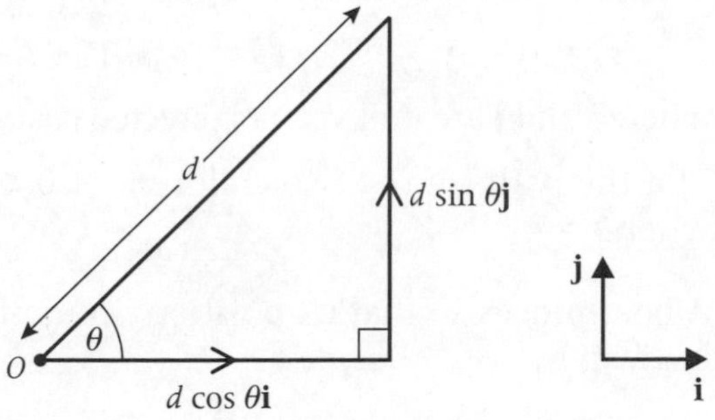

The following examples will illustrate how to write quantities in the form $a\mathbf{i} + b\mathbf{j}$.

Worked example 4

A ship starts at the origin and travels 200 km on a bearing of 140°. Express the position of the ship in the form $a\mathbf{i} + b\mathbf{j}$, where **i** and **j** are unit vectors that are directed east and north, respectively.

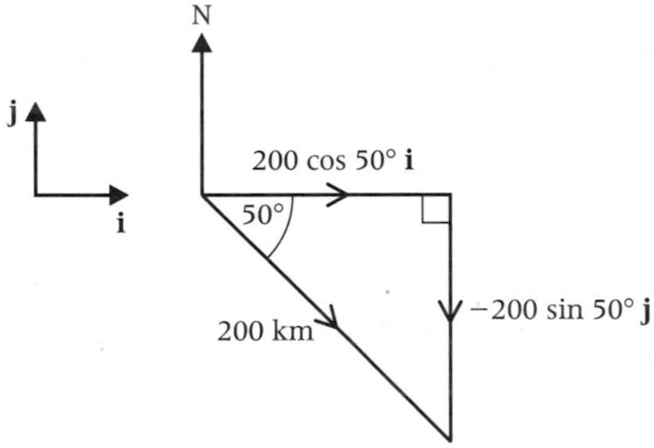

Solution

The diagram shows the position of the ship and the unit vectors.

In this case we can write:

$$\mathbf{r} = 200\cos 50°\mathbf{i} - 200\sin 50°\mathbf{j}$$

Worked example 5

A model aeroplane is travelling at 15 m s^{-1} on a bearing of 250°. Express the velocity of the aeroplane in the form $a\mathbf{i} + b\mathbf{j}$, where **i** and **j** are unit vectors that are directed east and north, respectively.

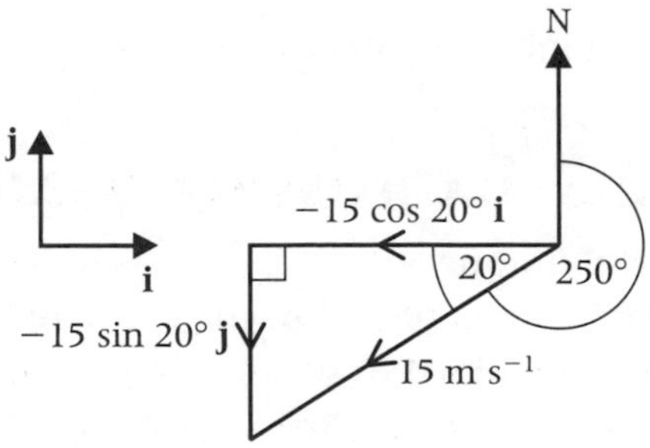

Solution

The diagram shows the velocity of the aeroplane and the unit vectors **i** and **j**. In this case the velocity of the aeroplane can be expressed as

$$\mathbf{v} = -15\cos 20°\mathbf{i} - 15\sin 20°\mathbf{j}$$

Worked example 6

The velocity of a bird is $3\mathbf{i} + 4\mathbf{j}$, where **i** and **j** are unit vectors directed east and north, respectively. Find the speed of the bird and the direction in which it is heading.

Solution

The diagram shows the velocity of the bird and the unit vectors. The speed of the bird is given by the magnitude or length of the velocity vector. This can be calculated using Pythagoras' theorem.

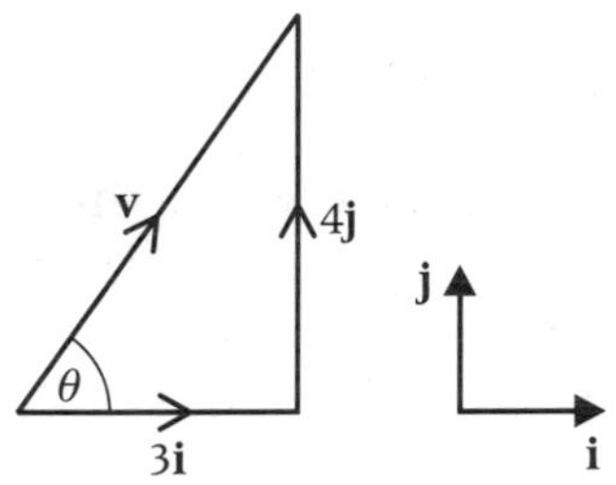

$$v = \sqrt{3^2 + 4^2}$$
$$= \sqrt{25}$$
$$= 5 \text{ m s}^{-1}$$

Next find the angle θ marked on the diagram.

$$\tan \theta = \frac{4}{3}$$

$$\theta = 53.1°$$

The best way to describe the direction of the velocity of the bird is by using the bearing of the direction that it is heading in. This is given by

$$90 - 53.1 = 036.9°$$

Note. The acceleration due to gravity is often expressed as $-g\mathbf{j}$, because it has magnitude g m s^{-2} and acts vertically downwards.

EXERCISE 3B

1 The unit vectors **i** and **j** are directed east and north respectively. The positions below are given in terms of a bearing and a distance. Express each position in the form $a\mathbf{i} + b\mathbf{j}$.

(a) 45 m on a bearing of 080°.

(b) 105 m on a bearing of 060°.

(c) 21 m on a bearing of 340°.

(d) 62 m on a bearing of 260°.

(e) 290 m on a bearing of 162°.

2 A boat sails south west at 6 m s^{-1}. Find the velocity of the ship in the form $a\mathbf{i} + b\mathbf{j}$, where **i** and **j** are unit vectors directed east and north, respectively.

3 A ship travels at a speed of $5\ \text{m s}^{-1}$. Express its velocity in terms of the unit vectors **i** and **j**, that are directed east and north, respectively, if the ship is sailing:

(a) due east

(b) due south

(c) due west

(d) south east

(e) north west.

4 A ball is thrown so that its initial velocity is $8\ \text{m s}^{-1}$ at an angle of 50° above the horizontal. The unit vectors **i** and **j** are horizontal and vertical, respectively. Express the initial velocity of the ball in terms of the unit vectors **i** and **j**.

5 For each velocity listed below, find its magnitude and direction. The unit vectors **i** and **j** are directed east and north, respectively. Give the directions as the bearing along which the velocity is directed.

(a) $(4\mathbf{i} + 7\mathbf{j})\ \text{m s}^{-1}$

(b) $(5\mathbf{i} - 6\mathbf{j})\ \text{m s}^{-1}$

(c) $(-8\mathbf{i} - 9\mathbf{j})\ \text{m s}^{-1}$

(d) $(-12\mathbf{i} + 8\mathbf{j})\ \text{m s}^{-1}$

6 An object moves along a straight line from the point with position vector $(5\mathbf{i} + 6\mathbf{j})$ m to the point with position vector $(8\mathbf{i} - 2\mathbf{j})$ m, where the unit vectors **i** and **j** are directed east and north, respectively.

(a) Find the distance travelled by the object.

(b) Find the bearing along which the object was travelling.

7 The acceleration of a particle is $(-3\mathbf{i} + 2\mathbf{j})\ \text{m s}^{-2}$, where the unit vectors **i** and **j** are directed north and east, respectively. Find the magnitude of the acceleration. Calculate the bearing along which the acceleration is directed.

3.4 Constant acceleration equations in two and three dimensions

The constant acceleration equations that were introduced and used in the last chapter can be extended into two and three dimensions as vector equations.

To make use of these vector equations the velocities, accelerations and positions must be in the form $a\mathbf{i} + b\mathbf{j}$ for two-dimensional motion, or $a\mathbf{i} + b\mathbf{j} + c\mathbf{k}$ for three dimensions.

The constant acceleration equations become

$$\mathbf{r} = \mathbf{u}t + \frac{1}{2}\mathbf{a}t^2 \quad \text{or} \quad \mathbf{r} = \mathbf{u}t + \frac{1}{2}\mathbf{a}t^2 + \mathbf{r}_0$$

$$\mathbf{v} = \mathbf{u} + \mathbf{a}t$$

$$\mathbf{r} = \frac{1}{2}(\mathbf{u} + \mathbf{v})t \quad \text{or} \quad \mathbf{r} = \frac{1}{2}(\mathbf{u} + \mathbf{v})t + \mathbf{r}_0$$

where **r** is the position at time t, **u** is the initial velocity, **v** is the velocity at time t, **a** is the acceleration and $\mathbf{r}_0$ is the initial position.

You will notice the similarity between these equations and the constant acceleration equations that you used in the last chapter. At this stage we will not use a vector equation that is equivalent to $v^2 = u^2 + 2as$.

The following examples illustrate how these formulae can be applied.

Worked example 7

A ball is rolling on an inclined plane. The initial velocity of the ball is $4\mathbf{i}$ m s^{-1}, its acceleration is $-2\mathbf{j}$ m s^{-2} and its initial position is $(9\mathbf{i} + 4\mathbf{j})$ m, where **i** and **j** are perpendicular unit vectors that lie in the plane on which the ball is moving.

(a) Find the velocity of the ball after it has been moving for 3 seconds.

(b) Find the position of the ball after it has been moving for 5 seconds.

Solution

For this problem:

$$\mathbf{u} = 4\mathbf{i}$$

$$\mathbf{a} = -2\mathbf{j}$$

$$\mathbf{r}_0 = 9\mathbf{i} + 4\mathbf{j}$$

(a) Using the formula for the velocity with the vectors above and $t = 3$ gives:

$$\begin{aligned}\mathbf{v} &= \mathbf{u} + \mathbf{a}t \\ &= 4\mathbf{i} + (-2\mathbf{j}) \times 3 \\ &= 4\mathbf{i} - 6\mathbf{j}\end{aligned}$$

(b) Using the formula for the position with the vectors listed above and $t = 5$ gives:

$$\mathbf{r} = \mathbf{u}t + \frac{1}{2}\mathbf{a}t^2 + \mathbf{r}_0$$

$$= 4\mathbf{i} \times 5 + \frac{1}{2}(-2\mathbf{j}) \times 5^2 + 9\mathbf{i} + 4\mathbf{j}$$

$$= 20\mathbf{i} - 25\mathbf{j} + 9\mathbf{i} + 4\mathbf{j}$$

$$= 29\mathbf{i} - 21\mathbf{j}$$

Worked example 8

A boat has initial velocity $(5\mathbf{i} + 3\mathbf{j})$ m s^{-1}. It accelerates for 4 seconds. Its velocity is then $(2\mathbf{i} - 4\mathbf{j})$ m s^{-1}. The boat then stops accelerating and travels with this velocity for a further 20 seconds. The initial position of the boat is $(24\mathbf{i} + 32\mathbf{j})$ m. The unit vectors **i** and **j** are directed east and north, respectively.

(a) Find the acceleration of the boat.

(b) Find the position of the boat after 4 seconds.

(c) Find the final position of the boat.

Solution

(a) The acceleration can be found using the equation $\mathbf{v} = \mathbf{u} + \mathbf{a}t$. Using $\mathbf{u} = 5\mathbf{i} + 3\mathbf{j}$, $\mathbf{v} = 2\mathbf{i} - 4\mathbf{j}$ and $t = 4$ gives:

$$\mathbf{v} = \mathbf{u} + \mathbf{a}t$$

$$2\mathbf{i} - 4\mathbf{j} = 5\mathbf{i} + 3\mathbf{j} + 4\mathbf{a}$$

$$4\mathbf{a} = -3\mathbf{i} - 7\mathbf{j}$$

$$\mathbf{a} = -\frac{3}{4}\mathbf{i} - \frac{7}{4}\mathbf{j}$$

(b) The position after 4 seconds can be found using the constant acceleration equation $\mathbf{r} = \frac{1}{2}(\mathbf{u} + \mathbf{v})t + \mathbf{r}_0$, with $\mathbf{u} = 5\mathbf{i} + 3\mathbf{j}$, $\mathbf{v} = 2\mathbf{i} - 4\mathbf{j}$, $\mathbf{r}_0 = 24\mathbf{i} + 32\mathbf{j}$ and $t = 4$.

$$\mathbf{r} = \frac{1}{2}(\mathbf{u} + \mathbf{v})\mathbf{t} + \mathbf{r}_0$$

$$= \frac{1}{2}(5\mathbf{i} + 3\mathbf{j} + 2\mathbf{i} - 4\mathbf{j}) \times 4 + 24\mathbf{i} + 32\mathbf{j}$$

$$= 14\mathbf{i} - 2\mathbf{j} + 24\mathbf{i} + 32\mathbf{j}$$

$$= 38\mathbf{i} + 30\mathbf{j}$$

(c) After 4 seconds the boat moves with a constant velocity for a further 20 seconds. As the acceleration is zero the equation $\mathbf{r} = \mathbf{u}t + \frac{1}{2}\mathbf{a}t^2 + \mathbf{r}_0$ reduces to $\mathbf{r} = \mathbf{u}t + \mathbf{r}_0$. This can be applied by considering only the motion after the boat stops accelerating. Using $\mathbf{u} = 2\mathbf{i} - 4\mathbf{j}$, $\mathbf{r}_0 = 38\mathbf{i} + 30\mathbf{j}$ and $t = 20$ gives

$$\begin{aligned}\mathbf{r} &= \mathbf{u}t + \mathbf{r}_0 \\ &= (2\mathbf{i} - 4\mathbf{j}) \times 20 + 38\mathbf{i} + 30\mathbf{j} \\ &= 78\mathbf{i} - 50\mathbf{j}\end{aligned}$$

Worked example 9

An aeroplane has a constant velocity of $120\mathbf{i}$ m s^{-1}, as it is moving along a runway. It then experiences an acceleration of $(2\mathbf{i} + 5\mathbf{j})$ m s^{-2} for the first 20 seconds of its flight. The unit vectors **i** and **j** are directed horizontally and vertically, respectively. Assume that the aeroplane is at the origin when it begins to accelerate.

(a) Find an expression for the position of the aeroplane at time t seconds, after it starts to accelerate.

(b) Find the speed of the aeroplane when it is at a height of 250 m.

Solution

(a) As the aeroplane is initially at the origin we can use the constant acceleration equation $\mathbf{r} = \mathbf{u}t + \frac{1}{2}\mathbf{a}t^2$, with $\mathbf{u} = 120\mathbf{i}$ and $\mathbf{a} = 2\mathbf{i} + 5\mathbf{j}$.

$$\begin{aligned}\mathbf{r} &= \mathbf{u}t + \frac{1}{2}\mathbf{a}t^2 \\ &= 120t\mathbf{i} + \frac{1}{2}(2\mathbf{i} + 5\mathbf{j})t^2 \\ &= (120t + t^2)\mathbf{i} + \frac{5}{2}t^2\mathbf{j}\end{aligned}$$

(b) The height of the aeroplane is given by the vertical or **j** component of the position vector. When the height of the aeroplane is 250 m we have:

$$\begin{aligned}\frac{5}{2}t^2 &= 250 \\ t^2 &= 100 \\ t &= 10 \text{ seconds}\end{aligned}$$

The velocity can now be found using this value for t.

$$\mathbf{v} = \mathbf{u} + \mathbf{a}t$$
$$= 120\mathbf{i} + (2\mathbf{i} + 5\mathbf{j}) \times 10$$
$$= 140\mathbf{i} + 50\mathbf{j}$$

The speed can be found as it will be the magnitude of the velocity.

$$v = \sqrt{140^2 + 50^2}$$
$$= \sqrt{22100}$$
$$= 149 \text{ m s}^{-1} \text{ (correct to three significant figures)}$$

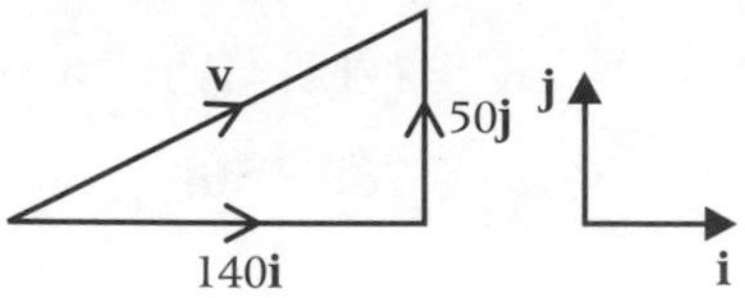

EXERCISE 3C

1 During a 10 second period the velocity of a boat changes from $(4\mathbf{i} + 2\mathbf{j})$ m s^{-1} to $(\mathbf{i} - 3\mathbf{j})$ m s^{-1}, where $\mathbf{i}$ and $\mathbf{j}$ are perpendicular unit vectors. Find the acceleration of the boat during this time, assuming that it is constant.

2 The acceleration of a motor boat is $(0.6\mathbf{i} + 0.8\mathbf{j})$ m s^{-2}. Its initial velocity is $3\mathbf{i}$ m s^{-1} and its initial position is $(20\mathbf{i} + 5\mathbf{j})$ m. The unit vectors, $\mathbf{i}$ and $\mathbf{j}$ are directed east and north, respectively.

(a) Find the velocity of the boat when $t = 3$ seconds.

(b) Find the position of the boat when $t = 3$ seconds.

3 An object has initial velocity $(3\mathbf{i} - 5\mathbf{j})$ m s^{-1} and an acceleration of $(\mathbf{i} + \mathbf{j})$ m s^{-2}, where $\mathbf{i}$ and $\mathbf{j}$ are perpendicular unit vectors. If it starts at the origin find the position and velocity at time t seconds.

4 The acceleration of a body is $(6\mathbf{i} + 8\mathbf{j})$ m s^{-2}. If the body starts at rest at $0\mathbf{i} + 0\mathbf{j}$ and accelerates for 6 seconds find the velocity and position of the body after 6 seconds. The unit vectors $\mathbf{i}$ and $\mathbf{j}$ are perpendicular.

5 A ball has initial position $2\mathbf{j}$ m, initial velocity $(4\mathbf{i} + 9\mathbf{j})$ m s^{-1} and acceleration $-10\mathbf{j}$ m s^{-2}, where $\mathbf{i}$ and $\mathbf{j}$ are horizontal and vertical unit vectors, respectively. Find the position of the ball at time t seconds and its position when it hits the ground, that is when the vertical component of its position is zero.

6 A snooker ball is struck so that it moves with the constant velocity shown in the diagram. It starts at the point with position vector $(0.2\mathbf{i} + 0.1\mathbf{j})$ m relative to the origin O. Find an expression for the position of the ball at time t, and find where it first hits a cushion. The unit vectors $\mathbf{i}$ and $\mathbf{j}$ are directed as shown on the diagram.

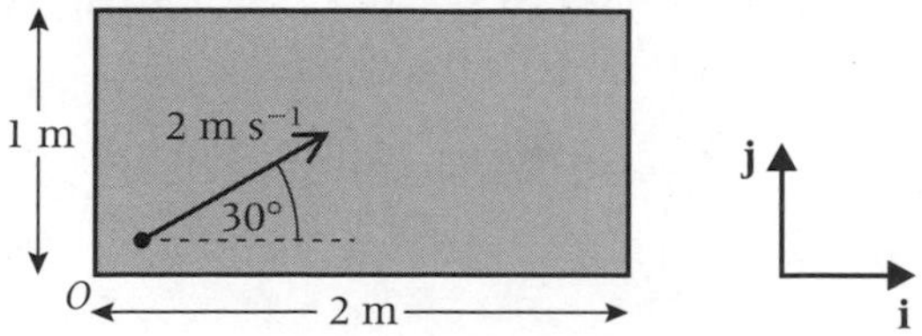

7 A ball is launched from a point with position $(0\mathbf{i} + 1.9\mathbf{j})$ m and velocity $(7\mathbf{i} + 32\mathbf{j})$ m s^{-1}. The ball experiences an acceleration of $(-9.8\mathbf{j})$ m s^{-2} while in the air. The unit vectors **i** and **j** are horizontal and vertical, respectively.

(a) Show that the position of the ball, in metres, at time t seconds is given by:

$$\mathbf{r} = 7t\mathbf{i} + (1.9 + 32t - 4.9t^2)\mathbf{j}.$$

(b) Find how long the ball is in the air and where it lands.

(c) Find the maximum height reached by the ball.

(d) Find the speed of the ball when it hits the ground.

8 A golf ball is hit from ground level so that its initial velocity is $(20\mathbf{i} + 30\mathbf{j})$ m s^{-1} and its acceleration is $-10\mathbf{j}$ m s^{-2}, where **i** and **j** are horizontal and vertical unit vectors, respectively. Assume that the initial position of the ball is $0\mathbf{i} + 0\mathbf{j}$.

(a) Find the time that the ball is in the air, the position of the point where it hits the ground and its speed at this time.

(b) Find the time for which the height of the ball is greater than 25 m.

9 The unit vectors **i** and **j**, are directed north and east, respectively. A boat has initial velocity $(4\mathbf{i} + 6\mathbf{j})$ m s^{-1} and an initial position of $(80\mathbf{i} + 20\mathbf{j})$ m. It experiences an acceleration of $(-0.02\mathbf{i} - 0.04\mathbf{j})$ m s^{-2} for a period of 4 minutes.

(a) Find an expression for the velocity and position of the particle at time t seconds.

(b) After 4 minutes the boat stops accelerating and continues with a constant velocity for a further minute before hitting the bank and stopping. Find the position of the point where the boat hits the bank.

(c) Find the velocity of the boat when its position is $476\mathbf{i} + 452\mathbf{j}$.

10 A model helicopter has initial position $80\mathbf{j}$ m and flies with a constant velocity of $(6\mathbf{i} - 3\mathbf{j})$ m s^{-1}. A model aeroplane flies at the same height as the helicopter. Its initial position is $10\mathbf{j}$ m, its initial velocity is $5\mathbf{i}$ m s^{-1} and its acceleration is $(0.1\mathbf{i} + 0.05\mathbf{j})$ m s^{-2}. The unit vectors **i** and **j** are directed east and north, respectively.

(a) Find expressions for the position vectors of the helicopter and the aeroplane at time t seconds.

(b) The helicopter and the aeroplane collide. Find the time when this collision takes place and the position of the collision.

11 The unit vectors **i** and **j**, are directed east and north, respectively. A boat has initial velocity $(2\mathbf{i} + 3\mathbf{j})$ m s^{-1} and an initial position of $(40\mathbf{i} + 20\mathbf{j})$ m with respect to an origin O. It experiences an acceleration of $(-0.06\mathbf{i} - 0.04\mathbf{j})$ m s^{-2}. Model the boat as a particle.

(a) Find expressions for the velocity and position, with respect to O, of the boat at time t seconds.

(b) The boat hits a sand bank when its position is $52\mathbf{i} + 128\mathbf{j}$. Find the value of t when this happens. [A]

12 A model assumes that an aeroplane has an initial velocity of $200\mathbf{i}$ m s^{-1} and experiences an acceleration of $(-0.5\mathbf{i} - 0.05\mathbf{j})$ m s^{-2} in preparation for landing. The initial position of the aeroplane was $(-50000\mathbf{i} + 4000\mathbf{j})$ m with respect to an origin O. The unit vectors **i** and **j** are horizontal and vertical, respectively. After accelerating for 200 seconds the velocity of the aeroplane is assumed to remain constant until it lands. The aeroplane lands when the vertical component of its position vector is zero.

(a) Find:

(i) The time it takes for the aeroplane to move from its initial position to the point where it first touches the ground.

(ii) The position of the aeroplane when it first touches the ground.

(iii) The speed of the aeroplane when it first touches the ground.

(b) Comment on the assumption that the velocity of the aeroplane is constant during the first stage of its flight. [A]

Key point summary

Formulae to learn

Constant acceleration equations for two or three dimensions

$$\mathbf{r} = \mathbf{u}t + \frac{1}{2}\mathbf{a}t^2 \quad \text{or} \quad \mathbf{r} = \mathbf{u}t + \frac{1}{2}\mathbf{a}t^2 + \mathbf{r}_0$$

$$\mathbf{v} = \mathbf{u} + \mathbf{a}t$$

$$\mathbf{r} = \frac{1}{2}(\mathbf{u} + \mathbf{v})t \quad \text{or} \quad \mathbf{r} = \frac{1}{2}(\mathbf{u} + \mathbf{v})t + \mathbf{r}_0$$

1 A position vector enables paths to be plotted and interpreted. *p29*

2 Positions, velocities and accelerations can be expressed in the form $x\mathbf{i} + y\mathbf{j} + z\mathbf{k}$ or $x\mathbf{i} + z\mathbf{j}$. *p29*

3 Magnitudes and directions of vectors can be found using trigonometric functions. *p34*

4 Constant acceleration equations can be used in two or three dimensions. *p37*

3

Test yourself

What to review

If your answer is incorrect – review

1 A ball moves so that its position, in metres, at time t seconds is given by

$$\mathbf{r} = 5t\mathbf{i} + (6 + 9.5t - 5t^2)\mathbf{j}$$

where the unit vectors **i** and **j** are horizontal and vertical, respectively. The origin is at ground level.

(a) Find the time when the ball hits the ground.

(b) Plot the path of the ball.

(c) Estimate the maximum height of the ball and the horizontal distance travelled by the ball from your plot.

Section 3.2

---

2 A ship travels at 3 m s^{-1} on a bearing of 235°. Express this velocity in the form $a\mathbf{i} + b\mathbf{j}$, where **i** and **j** are unit vectors that are directed east and north, respectively.

Section 3.3

---

3 The velocity of an aeroplane is $(80\mathbf{i} + 50\mathbf{j})$ m s^{-1}, where **i** and **j** are unit vectors that are directed east and north, respectively. Find the speed of the aeroplane and the direction in which it is heading.

Section 3.3

---

4 The unit vectors **i** and **j** are perpendicular and lie in a horizontal plane. A particle moves from the origin. Its initial velocity was $(4\mathbf{i} + 6\mathbf{j})$ m s^{-1} and after 20 seconds its velocity is $(24\mathbf{i} + 46\mathbf{j})$ m s^{-1}. The acceleration of the particle is constant.

(a) Find the acceleration of the particle.

(b) Find the distance of the particle from the origin after accelerating for 30 seconds.

Section 3.4

Test yourself ANSWERS

1 **(a)** 2.4 s, **(c)** 10.5 m, 12 m.

2 $-3\cos 35°\mathbf{i} - 3\sin 35°\mathbf{j}$.

3 94.3 m s^{-1}, 058.0°.

4 **(a)** $\mathbf{i} + 2\mathbf{j}$.

(b) 1221 m.

CHAPTER 4

Forces

Learning objectives

After studying this chapter you should be able to:

- identify forces acting on a body
- draw force diagrams
- resolve forces into components
- find resultant forces
- write forces as vectors
- understand that the resultant force is zero when the forces are in equilibrium
- know that for equilibrium a body must be at rest or moving with a constant speed
- use the friction inequality.

4.1 Introduction

In the earlier chapters of this book we have considered how to describe motion, using terms like velocity, acceleration and position. In this chapter we will consider forces. Forces cause motion or can act to keep objects at rest. An understanding of forces and how they cause motion is essential to be able to predict how object in real life will move.

> Force is a vector quantity – it has magnitude and direction. The unit of force is the newton, which is abbreviated to N.

We will look at the different types of force that can act on a body, and learn how to add them together to find their sum, which is called their resultant. We will investigate a particle in equilibrium. This is where the resultant force on the particle is zero, so that the particle remains at rest or moves with a constant velocity. Finally we will be looking at frictional forces. We will be required to find unknowns such as forces, angles, and so on. In order to do this it is essential that we are able to draw a clear force diagram, which shows all the forces acting on the particle.

4.2 Types of force

Weight

Weight is a force, which is the effect of the earth's gravitational pull. If this force acted on its own on a body it would cause it to accelerate. This acceleration is approximately 9.8 m s^{-2} (denoted by g). The force, which causes this acceleration, is mg N, where m is the mass of the particle, in kg.

Note that any two objects that are allowed to fall, will have the same acceleration and so fall at the same rate, even though the weight forces acting on them are different.

As forces are vectors, we will represent them on diagrams by arrows, which show the direction of the force. We will indicate the size or magnitude of the force by writing this next to the arrow. The diagram shows a force of magnitude F N and the direction in which it acts. In a sketch the length of the arrow is not important, but if you solve problems by scale drawing, then the length of the vector or arrow must be proportional to the magnitude of the force.

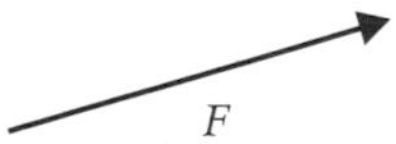

4

Particle on a plane

Suppose a particle with a weight of W N is at rest on a smooth horizontal plane. As the particle does not fall, the weight W is being balanced by an equal force acting upwards. This force is called the normal reaction, R, and acts perpendicular to the plane.

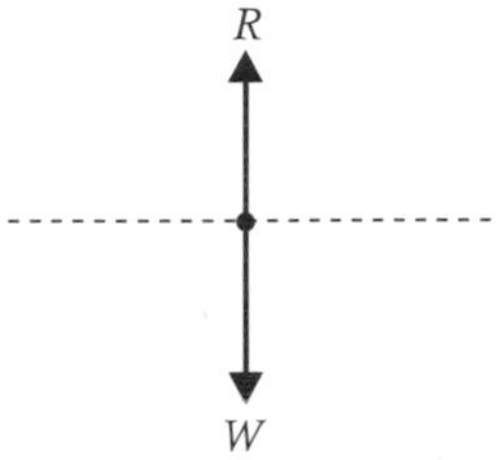

If we now place the particle on a rough horizontal plane, and pull the particle horizontally with a force of P N, then the roughness of the surface will oppose the motion. There will be a frictional force of F N, acting parallel to the plane, in the opposite direction. If the particle is at rest then $F = P$.

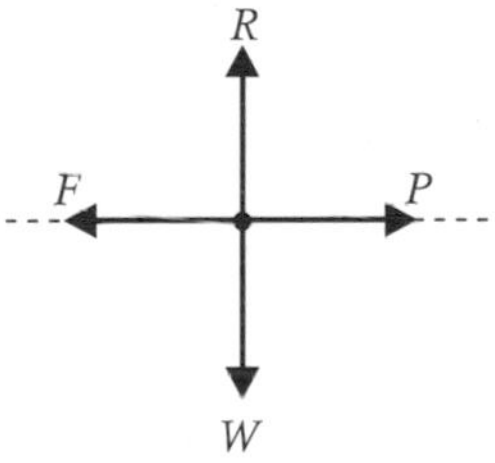

If we now consider the particle at rest on a rough plane, inclined at θ to the horizontal, then the frictional force will act up the slope, as shown in the diagram.

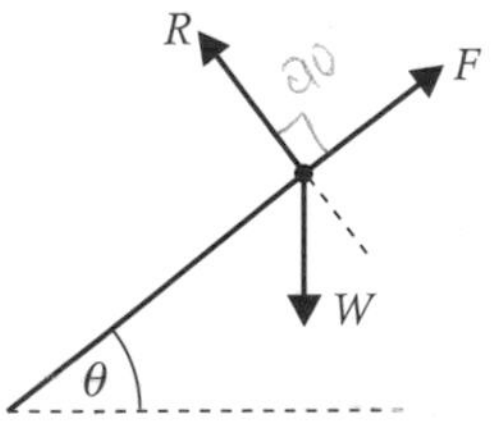

Strings and rods

A string will exert a force on an object if the string is taut, and this force, which is called a tension, will be directed along the string. For example, if a mass is suspended from a string, there will be a weight force acting downwards and a tension acting upwards.

Suppose a string is attached to a particle of weight W N, which rests on a smooth horizontal table. The only force, which the string can exert on the particle, will be in the same direction as the string itself. Furthermore, the string can only pull (and not push). Hence the tension, T, acts on the particle, as shown in the diagram.

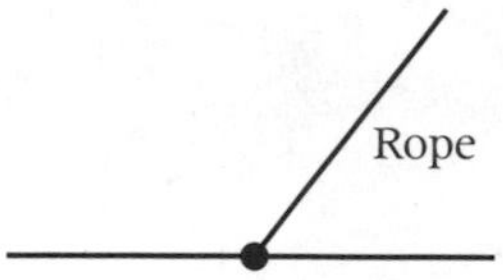

Situation

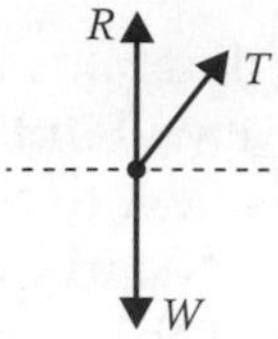

Force diagram

If a rod is attached to a particle instead of the string then the force of the rod on the particle could be in either of two directions, depending on whether it is pulling or pushing.

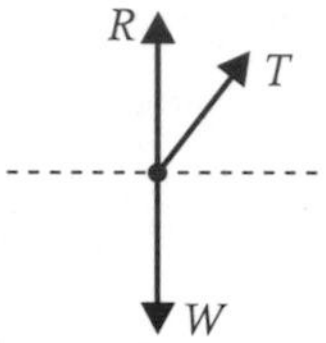

Rod pulling

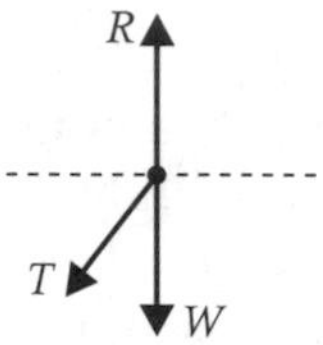

Rod pushing

Here are some examples where the force diagram has been drawn to demonstrate particular situations.

1 A particle of weight W N sliding down a smooth plane inclined at α to the horizontal. Note that the term smooth means that there is no friction.

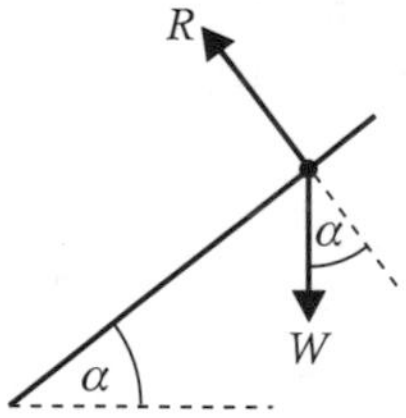

2 A particle of mass M kg at rest, suspended by a string.

3 A particle of mass M kg pulled across a rough horizontal plane by a string, inclined at 30° to the horizontal.

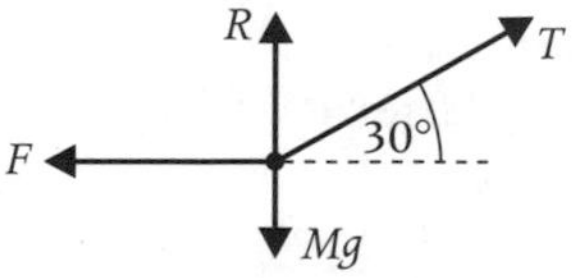

4 A and B are two points in the horizontal plane, a distance of 5 m apart. A particle, of weight W N, is attached by two strings, of length 3 m and 4 m, to the points A and B. The particle is at rest. The diagram shows the forces acting on the particle.

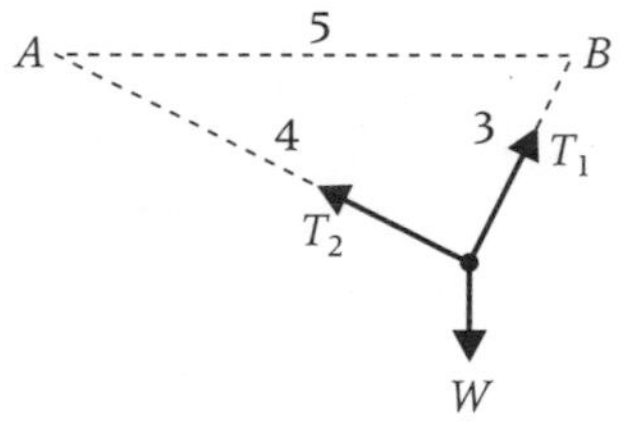

EXERCISE 4A

Draw force diagrams which show all the forces acting on the particle involved in each of the following situations:

1 A particle sliding down a rough plane inclined at α to the horizontal.

2 A particle of weight W N which is suspended from a fixed point by a string. The particle is held in equilibrium with the string at an angle θ to the vertical by a horizontal force P N.

3 A particle of mass m kg hangs in equilibrium supported by two light inextensible strings, inclined at 40° and 50° to the vertical.

4 A particle pushed up a rough plane, inclined at α to the horizontal, by a light horizontal rod, that is itself at an angle β to the slope.

5 A particle of weight W N is attached to one end of a light rod. The other end of the rod is fixed. A horizontal force P pulls the particle sideways, so that the rod makes an angle of 20° with the vertical.

4.3 Resultant forces

Adding two forces

Consider two forces, of magnitude F_1 and F_2, which act upon a particle. If we place these forces end to end, it can be seen that they have the same effect as a single force, of magnitude R. This force is known as the resultant force.

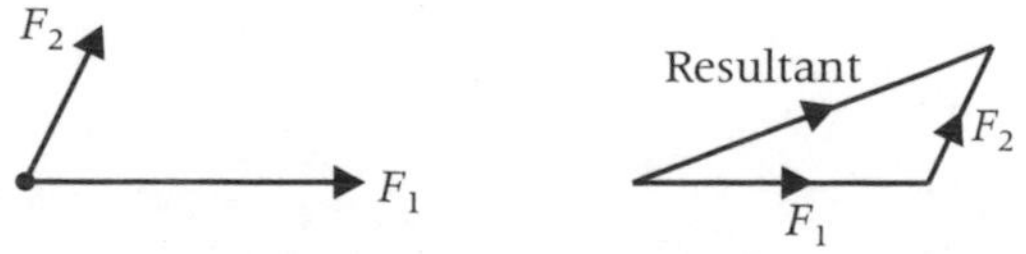

The resultant force will form the third side in a triangle of forces.

Worked example 1

Two forces of magnitudes 6 N and 5 N, act on a particle. The angle between the forces is 40°. Find the magnitude and direction of the resultant force.

Solution

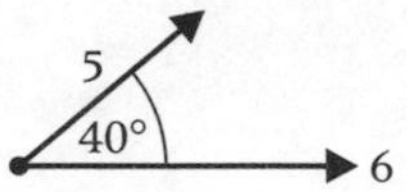

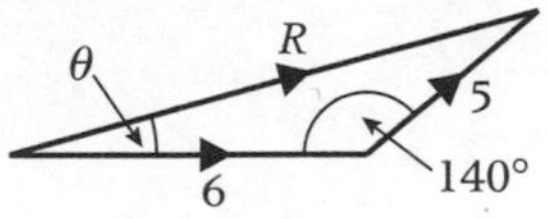

Using the cosine rule in the triangle of forces:

$$R^2 = 6^2 + 5^2 - 2 \times 6 \times 5 \times \cos 140°$$

$$R = 10.3 \text{ N}$$

Applying the sine rule:

$$\frac{\sin \theta}{5} = \frac{\sin 140}{R}$$

$$\theta = 18.1°$$

So the resultant force has magnitude 10.3 N and is at 18.1° to the 6 N force.

Worked example 2

Forces, of magnitude 14 N and 8 N, act on a particle. The resultant force has magnitude 17 N. Find the angle between the forces.

Solution

The required angle is θ, as shown. However, it is easier to first find the angle marked x, in the triangle of forces.

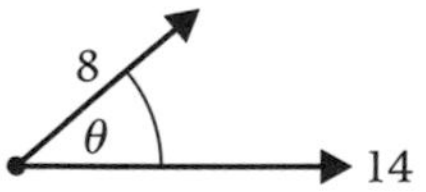

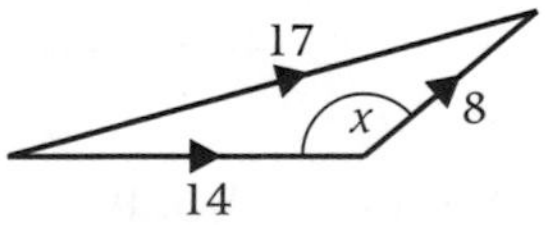

Using the cosine rule:

$$\cos x = \frac{14^2 + 8^2 - 17^2}{2 \times 14 \times 8}$$

$$x = 97.4°$$

The angle, θ, between the forces is:

$$180 - 97.4 = 82.6°$$

EXERCISE 4B

1 Find the magnitude of the resultant of the forces, of magnitude F_1 and F_2, if θ is the angle between them. Also find the angle between the resultant and the force of magnitude F_1.

(a) $F_1 = 4$ N, $F_2 = 3$ N, $\theta = 90°$

(b) $F_1 = 6$ N, $F_2 = 10$ N, $\theta = 60°$

(c) $F_1 = 7$ N, $F_2 = 9$ N, $\theta = 75°$

(d) $F_1 = 8$ N, $F_2 = 12$ N, $\theta = 170°$

(e) $F_1 = 10$ N, $F_2 = 11$ N, $\theta = 160°$

2 Forces, of magnitude F_1 and F_2, act on a particle. The resultant of the forces has magnitude R. Find the angle between the two forces acting on the particle in the following cases:

(a) $F_1 = 60$ N, $F_2 = 80$ N, $R = 100$ N

(b) $F_1 = 11$ N, $F_2 = 17$ N, $R = 20$ N

(c) $F_1 = 7$ N, $F_2 = 10$ N, $R = 5$ N

(d) $F_1 = 8$ N, $F_2 = 7$ N, $R = 3$ N

3 Forces of magnitude 6 N and 5 N act on a particle. What are the greatest and least values of the magnitude of the resultant of these forces?

4 The resultant of two forces, of magnitude F_1 and F_2, has magnitude 20 N. If $F_1 = 8$ N and the angle between the two forces is 70°, find F_2.

5 The resultant of two forces, of magnitude P and Q has magnitude 15 N. If $Q = 6$ N and the angle between the two forces is 60°, find P.

6 The resultant of two forces, of magnitude F_1 and F_2 has magnitude 3 N. If $F_1 = 7$ N and $F_2 = 5$ N, calculate the angle between the two forces.

7 The resultant of two forces, of magnitude F_1 and F_2, has magnitude 12 N, and acts at an angle of 30° to F_1. If $F_1 = 6$ N find the magnitude of F_2 and direction of F_2.

4.4 Adding any number of forces

Three forces can be added by placing one force on the end of the other. The resultant of the forces is represented by the vector that can be drawn along the fourth side of the resulting quadrilateral, as shown below. The magnitude of the resultant, R, is given by the length of the fourth side.

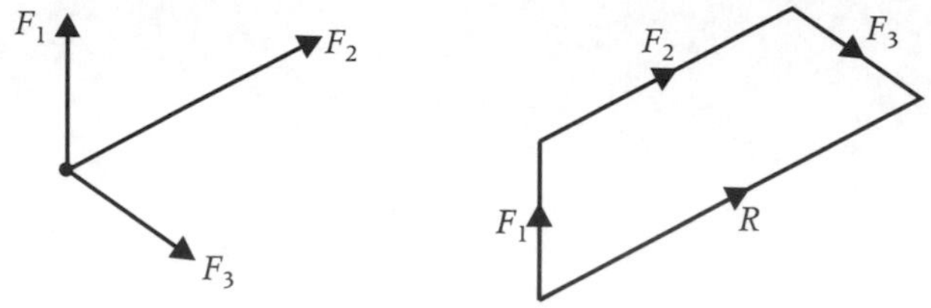

The quadrilateral can be divided into two triangles. The sine and cosine rules can then be used in the triangles to find the magnitude and direction of the resultant force.

Worked example 3

Find the magnitude of the resultant of the following set of forces:

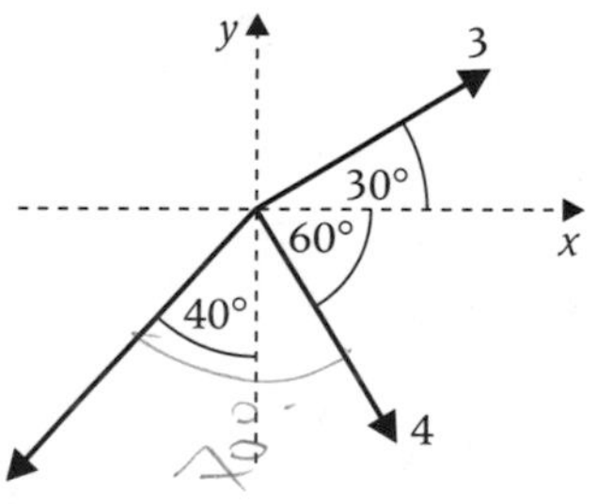

Solution

The diagram shows the three forces placed end to end to form a quadrilateral. The magnitude of the resultant force is R.

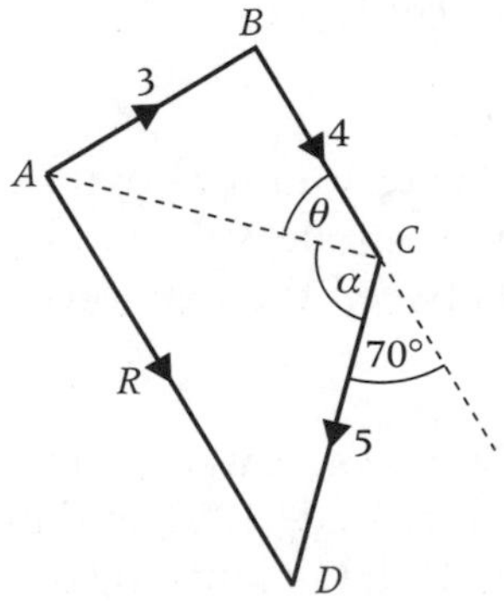

The angle between the 3 N force and the 4 N force is 90°. Therefore Pythagoras' theorem can be used to find AC in triangle ABC.

$$AC^2 = 3^2 + 4^2,$$

$$AC = 5.$$

Also $\tan\theta = \frac{3}{4}$

$$\theta = 36.9°.$$

$$\alpha = 180 - \theta - 70 = 73.1°$$

Using the cosine rule in triangle *ACD* gives:

$$R^2 = 5^2 + 5^2 - 2 \times 5 \times 5 \cos 73.1°$$

$$R = 5.96 \text{ N}$$

It is also possible to show that the angle between *R* and the *x*-axis is 76.6°.

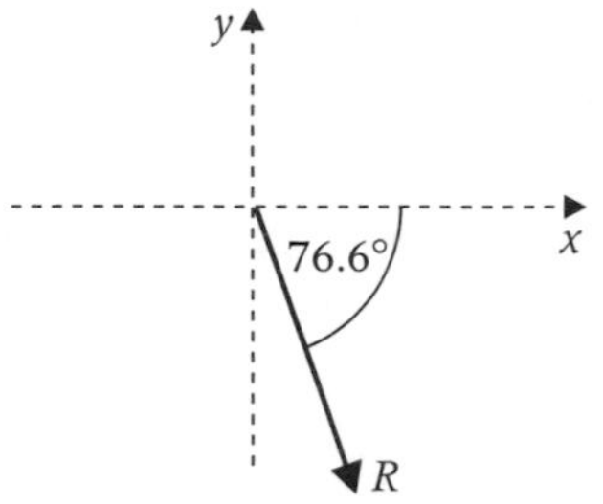

Using forces given in component form

> The approach of Worked example 3 is a slow and complicated process in general. The resultant of a set of forces can be found more quickly when the forces are given in component form.

In the same way that velocities and other vectors were expressed in the form $a\mathbf{i} + b\mathbf{j}$ in an earlier chapter, forces can be expressed in the same way. For example a force that has magnitude 40 N and acts at 30° above the horizontal could be expressed as:

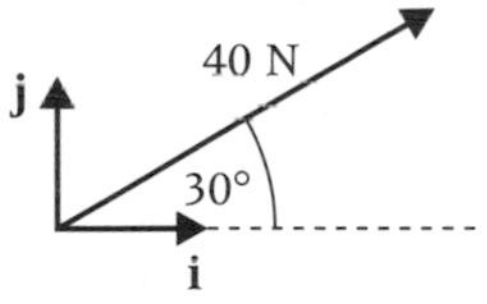

$$40 \cos 30°\mathbf{i} + 40 \sin 30°\mathbf{j}$$

First, we will concentrate on working with forces that are already expressed in vector form.

Worked example 4

Find the magnitude of the resultant of the following sets of forces:

$$\mathbf{F}_1 = (2\mathbf{i} + 3\mathbf{j} + \mathbf{k}) \text{ N}, \mathbf{F}_2 = (\mathbf{i} - 2\mathbf{j} + 4\mathbf{k}) \text{ N}, \mathbf{F}_3 = (2\mathbf{j} - 6\mathbf{k}) \text{ N}.$$

Solution

First find the resultant in terms of **i** and **j**.

$$\mathbf{R} = \mathbf{F}_1 + \mathbf{F}_2 + \mathbf{F}_3$$
$$= 2\mathbf{i} + 3\mathbf{j} + \mathbf{k} + \mathbf{i} - 2\mathbf{j} + 4\mathbf{k} + 2\mathbf{j} - 6\mathbf{k}$$
$$= 3\mathbf{i} + 3\mathbf{j} - \mathbf{k}$$

Then find the magnitude of the force.

$$R = \sqrt{3^2 + 3^3 + (-1)^2}$$
$$= \sqrt{19} = 4.36 \text{ N (to three significant figures)}$$

Worked example 5

The forces $(5\mathbf{i} + 12\mathbf{j})$ N and $(2\mathbf{i} + 4\mathbf{j})$ N act at a point.

(a) Find the magnitude of the resultant force.

(b) Find the angle between the resultant force and the unit vector $\mathbf{i}$.

Solution

(a) Resultant $= (5\mathbf{i} + 12\mathbf{j}) + (2\mathbf{i} + 4\mathbf{j})$

$= 7\mathbf{i} + 16\mathbf{j}$

Magnitude $= \sqrt{7^2 + 16^2}$

$= 17.5$ N to three significant figures

(b) Angle $= \tan^{-1}\left(\dfrac{16}{7}\right)$

$= 66.4°$

EXERCISE 4C

1 Find the magnitude of the resultant of the following sets of forces, by forming a quadrilateral of forces.

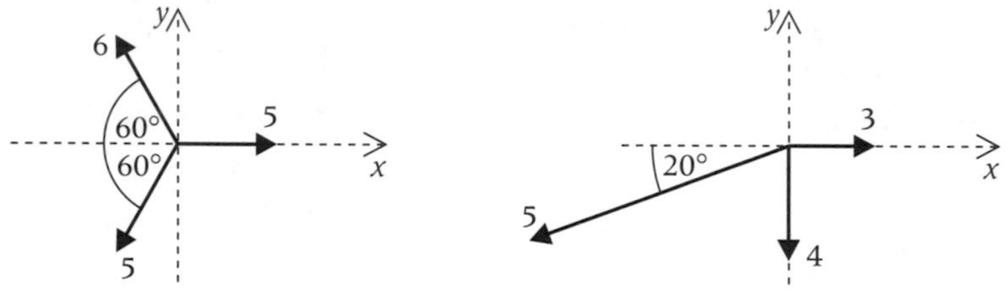

(a) **(b)**

2 Find the magnitude of the resultant of the forces $(2\mathbf{i} + \mathbf{j} + \mathbf{k})$ N, $(3\mathbf{i} - 2\mathbf{j} + 3\mathbf{k})$ N and $(-2\mathbf{i} + \mathbf{k})$ N.

3 The resultant of the forces $(2\mathbf{i} + \mathbf{j})$ N, $3\mathbf{j}$ N, $(2\mathbf{i} + 4\mathbf{j})$ N, $(6\mathbf{i} + b\mathbf{j})$ N and $(a\mathbf{i} + \mathbf{j})$ N is $(3\mathbf{i} + 4\mathbf{j})$ N, where $\mathbf{i}$ and $\mathbf{j}$ are perpendicular unit vectors. Find a and b.

4 Three forces $(3\mathbf{i} + 5\mathbf{j})$ N, $(4\mathbf{i} + 11\mathbf{j})$ N, $(2\mathbf{i} + \mathbf{j})$ N act at a point. Given that $\mathbf{i}$ and $\mathbf{j}$ are perpendicular unit vectors find:

(a) the resultant of the forces in the form $a\mathbf{i} + b\mathbf{j}$,

(b) the magnitude of this resultant,

(c) the angle that the resultant makes with the unit vector $\mathbf{i}$. [A]

5 Two forces $(3\mathbf{i} + 2\mathbf{j})$ N and $(-5\mathbf{i} + \mathbf{j})$ N act at a point. Find the magnitude of the resultant of these forces and determine the angle which the resultant makes with the unit vector $\mathbf{i}$. [A]

6 Three forces $(\mathbf{i}+\mathbf{j})$ N, $(-5\mathbf{i}+3\mathbf{j})$ N and $\lambda\mathbf{i}$ N, where **i** and **j** are perpendicular unit vectors, act at a point. Express the resultant in the form $(a\mathbf{i}+b\mathbf{j})$ and find its magnitude in terms of λ. Given that the resultant has magnitude 5 N, find the two possible values of λ.

Take the larger value of λ and find the tangent of the angle between the resultant and the unit vector **i**. [A]

4.5 Resolving forces

A force can be divided into two mutually perpendicular components whose vector sum is equal to the given force.

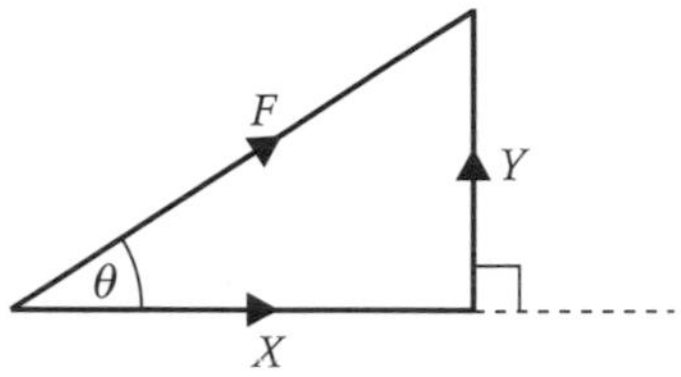

From the right-angled triangle:

$$X = F\cos\theta$$

and

$$Y = F\sin\theta$$

This process enables us to write forces in the form $a\mathbf{i} + b\mathbf{j}$. In the case above we would write:

$$F\cos\theta\mathbf{i} + F\sin\theta\mathbf{j}$$

Worked example 6

A force of 10 N acts at 60° below the horizontal. The unit vectors **i** and **j** are horizontal and vertical, respectively. Write the force in terms of the unit vectors **i**, and **j**.

Solution

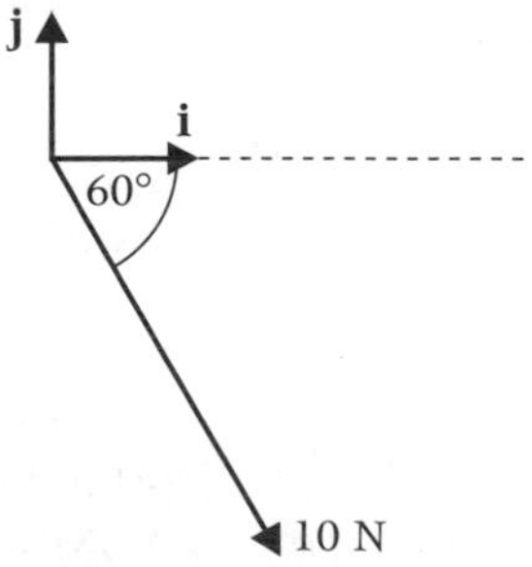

The diagram shows the force and the unit vectors.

The horizontal component of the force is 10 cos 60°.

The vertical component of the force is −10 sin 60°

Hence the force can be written as:

$$10\cos 60°\mathbf{i} - 10\sin 60°\mathbf{j} = 5\mathbf{i} - 8.66\mathbf{j} \text{ (to three significant figures)}$$

Worked example 7

Find the magnitude and direction of the resultant of the set of forces given below.

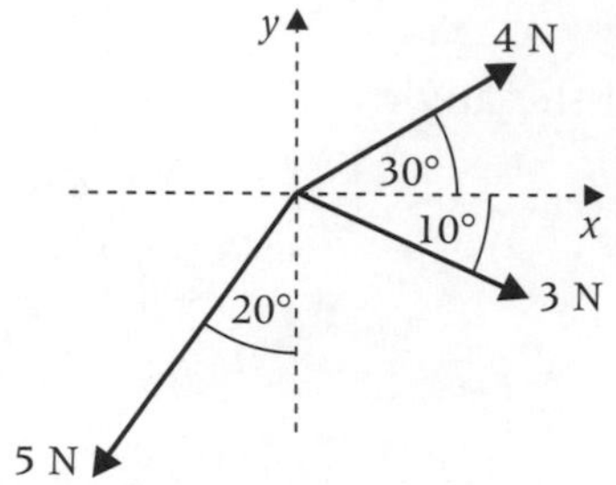

Solution

Resolving horizontally:

$$4\cos 30° + 3\cos 10° - 5\sin 20° = 4.708$$

Resolving vertically:

$$4\sin 30° - 3\sin 10° - 5\cos 20° = -3.219$$

The resultant force has magnitude $R = \sqrt{4.708^2 + 3.219^2} = 5.70$ N (to three significant figures) and acts at an angle

$\tan^{-1}\left(\frac{3.219}{4.708}\right) = 34.4°$ below the positive x-axis.

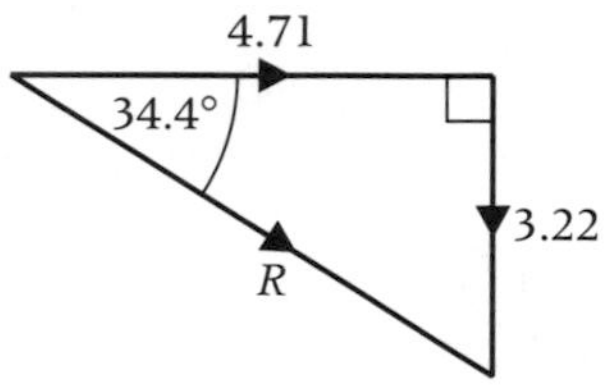

Examiner's tip: It is a good idea to draw a diagram in such cases, as it makes your intention clear

EXERCISE 4D

1 Find the components of the following forces in the directions of the x axis and y axis.

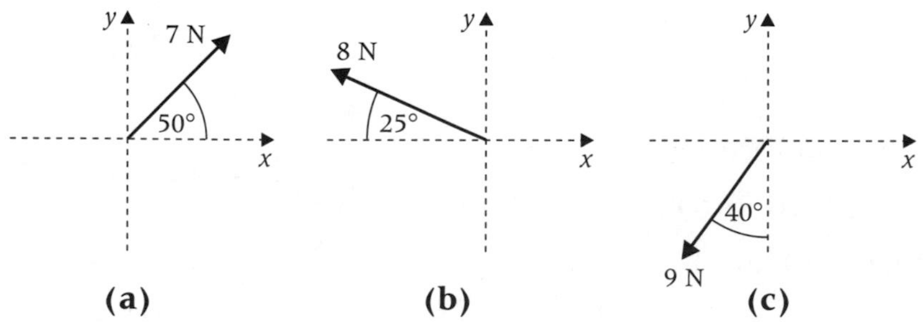

2 Express the forces in question 1 in the form $a\mathbf{i} + b\mathbf{j}$, if $\mathbf{i}$ is directed along the x axis and $\mathbf{j}$ is directed along the y axis.

3 Find the components of the following forces in the directions of the x axis and y axis.

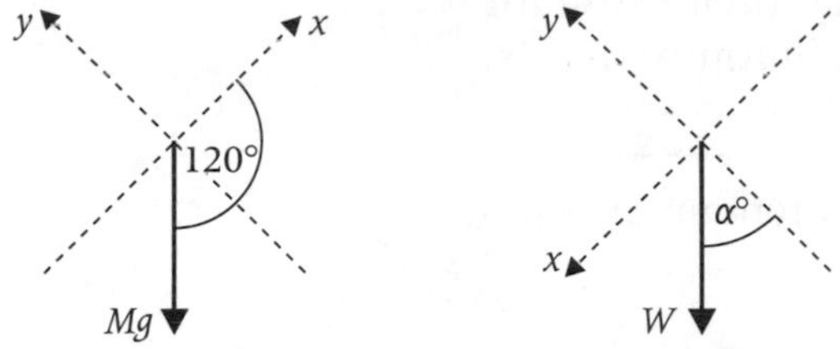

(a) **(b)**

4 Find the magnitude and the direction of the resultant force in each of the following cases:

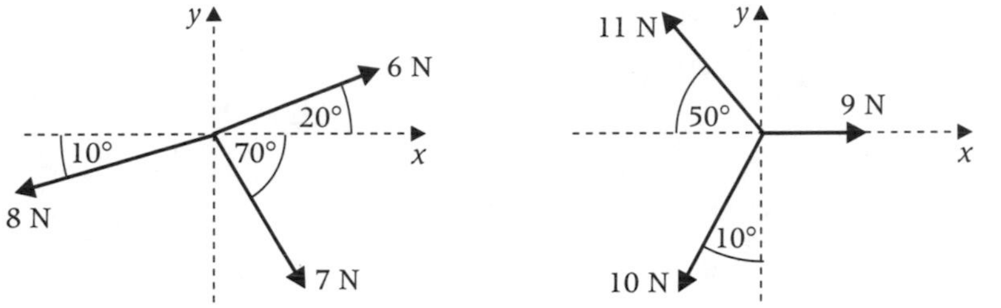

(a) **(b)**

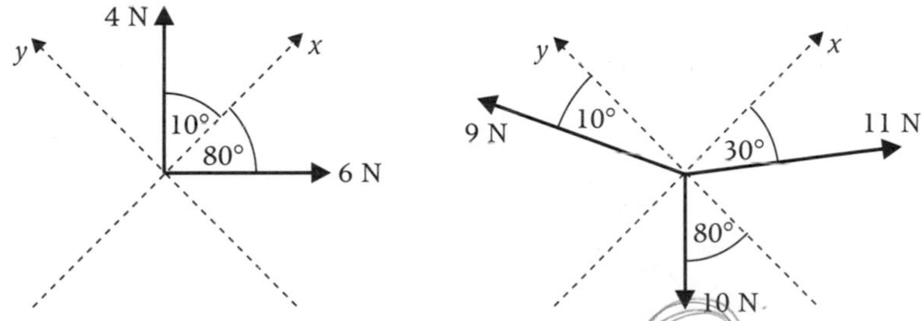

(c) **(d)**

5 Three horizontal forces each of magnitude 10 N act in the directions of the bearings 040°, 160° and 280°. Find the magnitude of their resultant.

6 The following diagrams show a particle on an inclined plane. Find the components of the weight of the particle in the directions of the x axis and y axis.

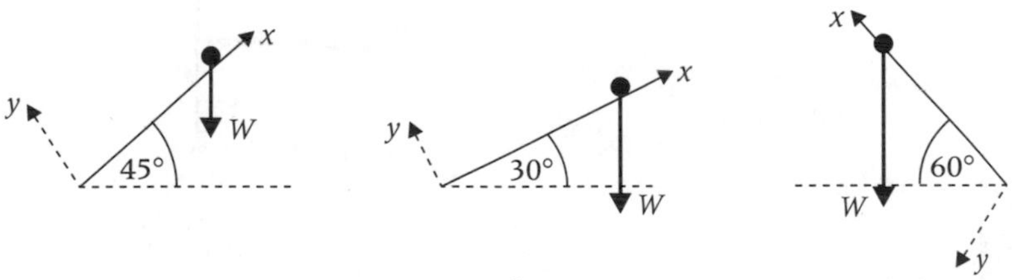

(a) **(b)** **(c)**

7 A particle of mass m kg lies at rest on a rough plane, inclined at α to the horizontal. What is the component of the weight down the plane?

4.6 Equilibrium

> If a set of forces act on a particle such that their resultant is zero, then the particle is said to be in equilibrium: that is, no unbalanced forces act on the particle.

When a particle is in equilibrium it will either remain at rest or move with a constant velocity.

Two forces

If only two forces act on a particle, which is in equilibrium, then the forces must be equal and opposite. For example, a particle of mass m kg rests in equilibrium, suspended by a light inextensible string.

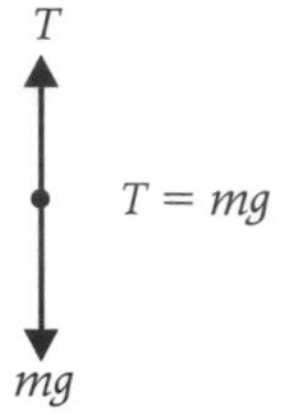

Three forces

If forces, of magnitude F_1, F_2 and F_2, have a zero resultant, then when the forces are placed end to end they must make a triangle.

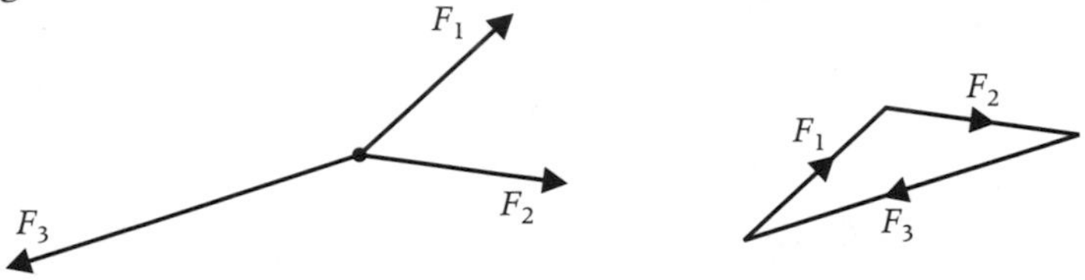

The sine rule and the cosine rule can be used in this triangle.

Worked example 8

A particle of mass 10 kg hangs in equilibrium suspended by two light inextensible strings, as shown. Find the tensions in the strings.

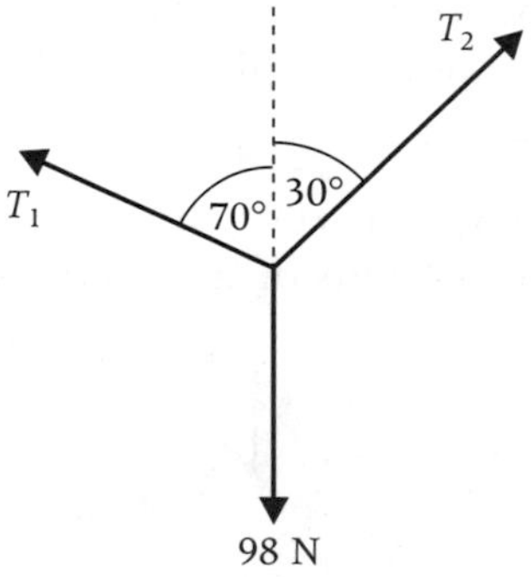

Solution

The diagram below shows how the forces must form a triangle, as they are in equilibrium.

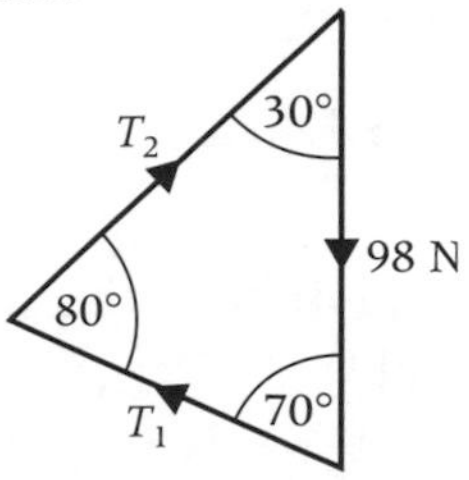

From the sine rule:

$$\frac{T_1}{\sin 30°} = \frac{T_2}{\sin 70°} = \frac{98}{\sin 80°}$$

$$T_1 = \frac{98 \sin 30°}{\sin 80°} \quad \text{and} \quad T_2 = \frac{98 \sin 70°}{\sin 80°}$$

$T_1 = 49.8$ N, $T_2 = 93.5$ N (to three significant figures)

Worked example 9

The set of forces shown below is in equilibrium. Find P and θ.

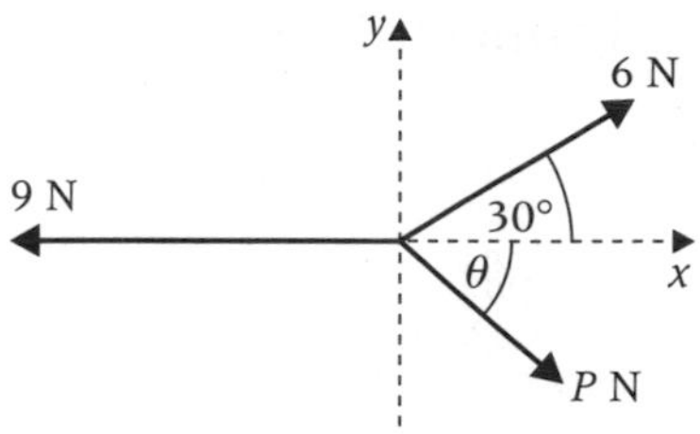

Solution

As the forces are in equilibrium, they can be arranged to form a triangle as shown in the diagram.

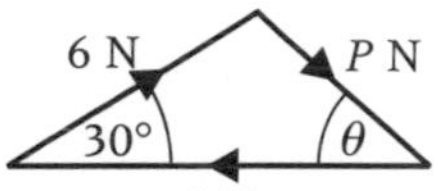

Using the cosine rule in this triangle gives:

$$P^2 = 6^2 + 9^2 - 2 \times 6 \times 9 \times \cos 30°$$

$$P = 4.84 \text{ N}$$

From the sine rule:

$$\frac{\sin \theta}{6} = \frac{\sin 30°}{4.84}$$

$$\theta = 38.3°$$

More than three forces

The previous two problems were solved using the sine and cosine rules only. Alternative solutions can be produced by resolving the forces in two perpendicular directions, for example horizontal and vertical. This method is particularly useful when we have more than three forces acting on the particle.

Worked example 10

The following set of forces is in equilibrium. Find P and the angle θ.

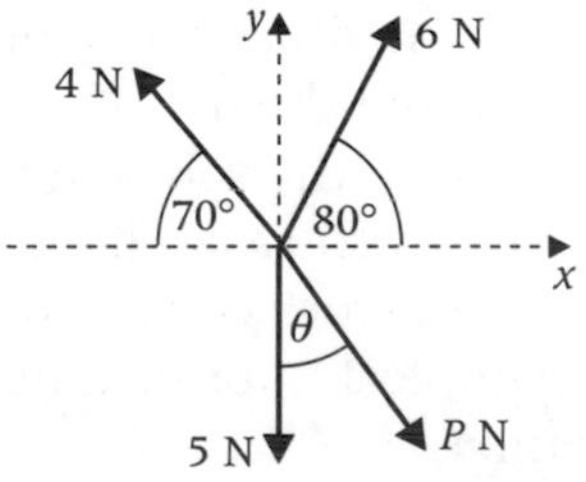

Solution

Resolving vertically:

$P \cos \theta + 5 = 4 \sin 70° + 6 \sin 80°$ (1)

$P \cos \theta = 4.667$

Resolving horizontally:

$P \sin \theta + 6 \cos 80° = 4 \cos 70°$ (2)

$P \sin \theta = 0.326$

Simultaneous equations like (1) and (2) occur frequently in mechanics. They can be solved using the trigonometric identities

$$\text{“}\sin^2 \theta + \cos^2 \theta \equiv 1\text{”} \qquad \text{“}\tan \theta \equiv \frac{\sin \theta}{\cos \theta}\text{”}$$

Hence $\tan \theta = \dfrac{0.326}{4.667}$

$\theta = 4.0°$

and $P^2 = 4.667^2 + 0.326^2$

$P = 4.68$ N

Worked example 11

A particle of mass 5 kg is at rest on a rough plane inclined at 50° to the horizontal. Find the normal reaction and the frictional force on the particle.

Solution

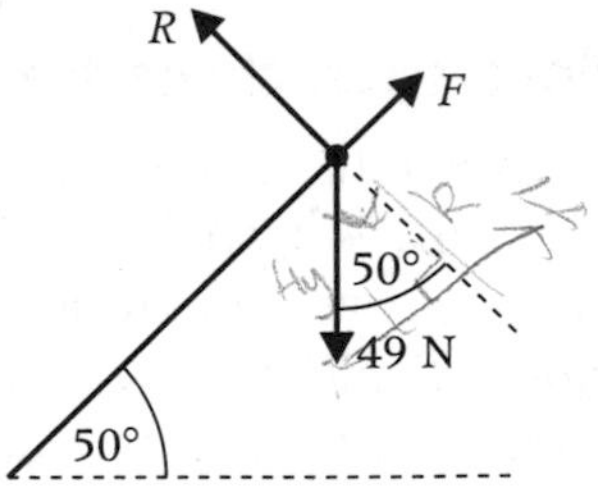

The diagram shows the forces acting on the particle.

Resolving parallel to the plane:

$F = 49 \sin 50° = 37.5$ N

Resolving perpendicular to the plane:

$R = 49 \cos 50° = 31.5$ N

Examiner's tip: The choice of whether to solve a problem by resolving or by drawing the triangle of forces and using trigonometry depends upon the particular situation involved. In the preceding examples, two demonstrated the use of trigonometry, and two showed how to find the solution by resolving. Whereas some problems are much easier done one way than the other, either technique will work equally well for certain types of problem. Worked example 11, for instance could be solved equally well by either method.

EXERCISE 4E

1 Each of the following sets of forces is in equilibrium. Find the magnitudes F_1 and F_2.

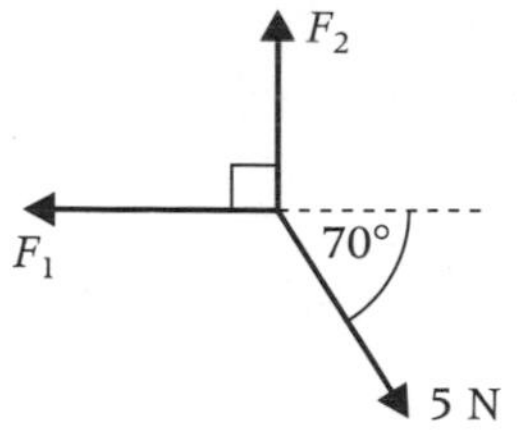

(a)

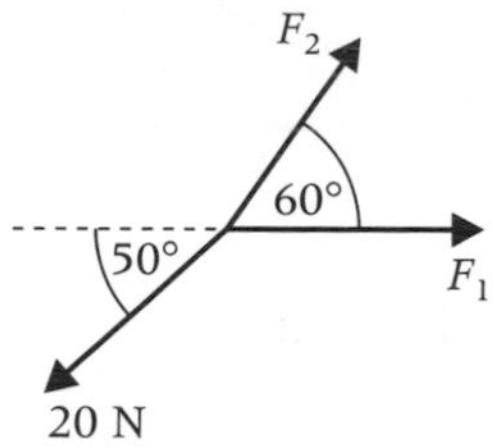

(b)

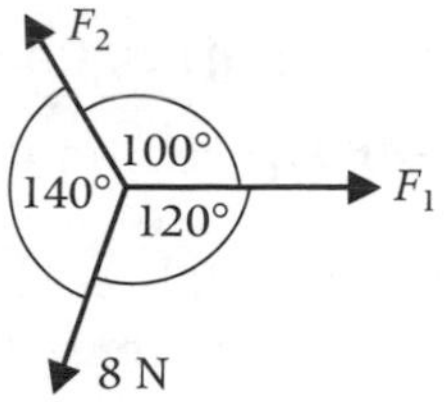

(c)

2 Each of the following sets of forces is in equilibrium. Find F and the angle θ.

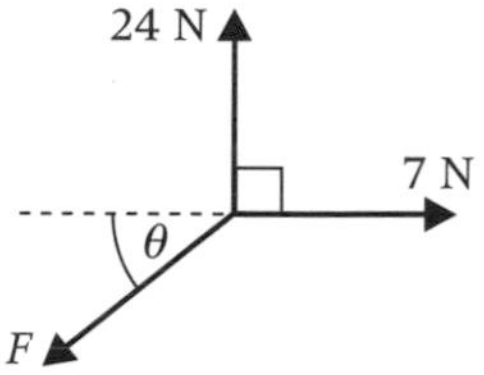

(a)

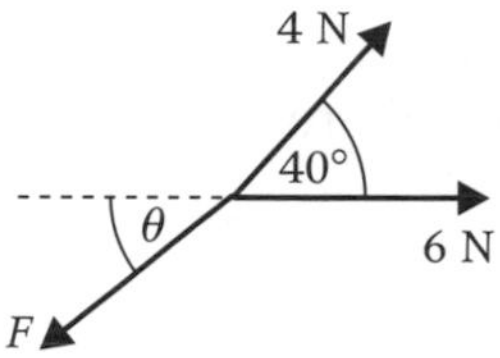

(b)

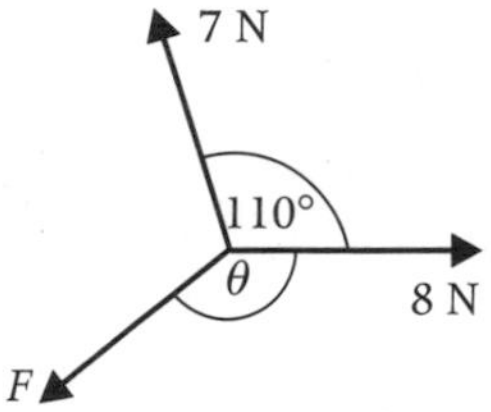

(c)

3 Each of the following sets of forces is in equilibrium. Find F_1 and F_2.

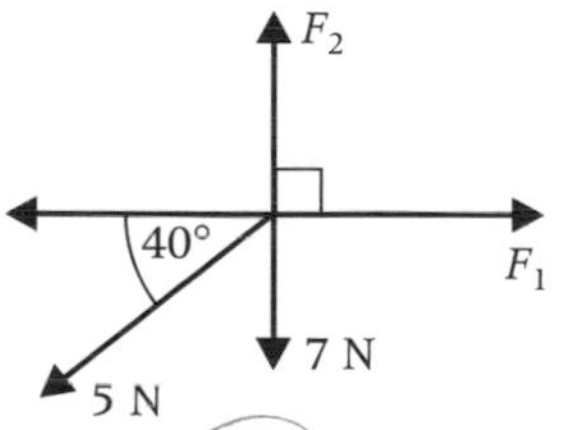

(a)

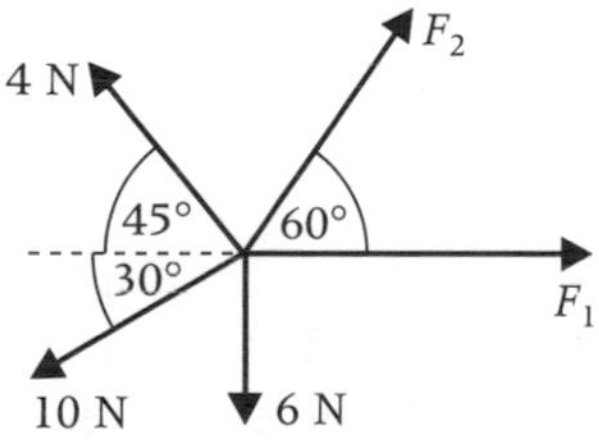

(b)

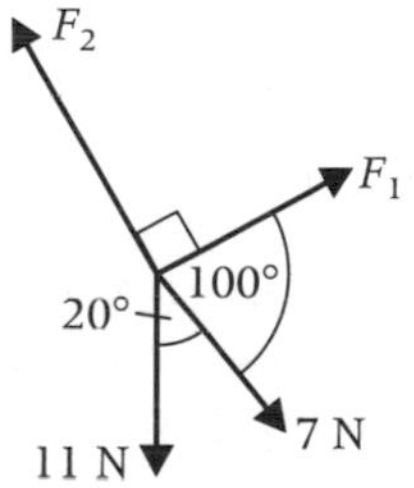

(c)

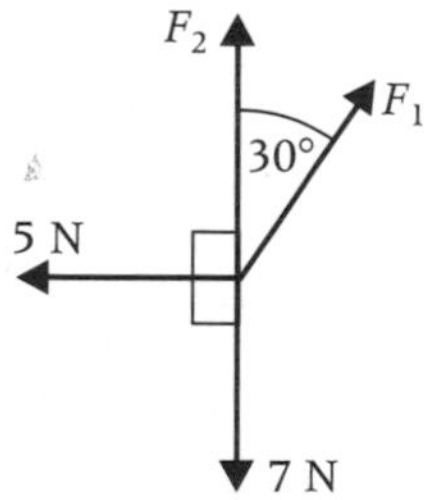

(d)

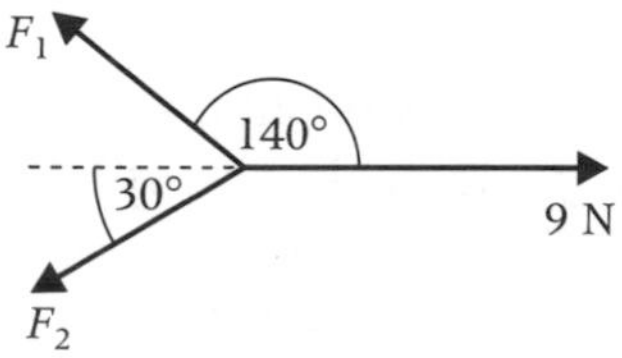

(e)

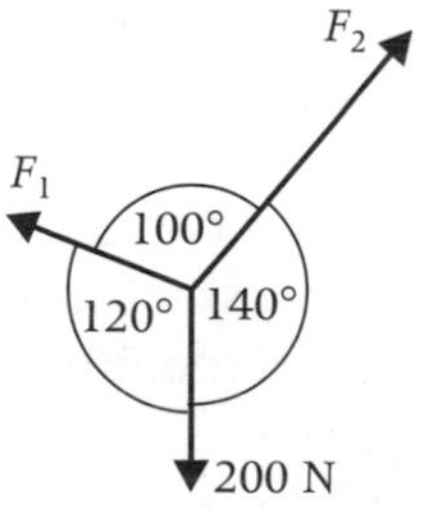

(f)

4 A particle is in equilibrium, subject to forces $(6\mathbf{i} + \mathbf{j})$ N, $(2\mathbf{i} + 3\mathbf{j})$ N and $\mathbf{P}$.

(a) Find $\mathbf{P}$ in terms of $\mathbf{i}$ and $\mathbf{j}$.

(b) Find the magnitude of $\mathbf{P}$ and the angle between $\mathbf{P}$ and $\mathbf{i}$.

5 The set of forces $(a\mathbf{i} + 2\mathbf{j})$ N, $(6\mathbf{i} - 3\mathbf{j})$ N, $(4\mathbf{i} + b\mathbf{j})$ N, and $(\mathbf{i} + \mathbf{j})$ N are in equilibrium. Find a and b.

6 Forces $\mathbf{F}_1$, $\mathbf{F}_2$ and $\mathbf{F}_3$ are in equilibrium.

(a) If $\mathbf{F}_1 = (\mathbf{i} + \mathbf{j})$ N, and $\mathbf{F}_3 = (7\mathbf{i} + 8\mathbf{j})$ N, find $\mathbf{F}_2$.

(b) If $\mathbf{F}_1 = (2\mathbf{i} + \mathbf{j})$ N, $\mathbf{F}_2 = (-4\mathbf{i} + \mathbf{j})$ N, find $\mathbf{F}_3$.

7 Horizontal forces each of magnitude 10 N act in the direction of the bearings 040°, 160° and 280°. Are these forces in equilibrium?

8 A mass of 5 kg is suspended by two light, inextensible strings. The angles between the strings and the vertical are 30° and 60°. Find the tensions in the strings.

9 A mass of 10 kg is suspended by two light, inextensible strings. The angles between the strings and the vertical are 40° and 20°. Find the tensions in the strings.

10 A particle of weight 10 N is suspended by two light, inextensible strings. The tension in one string is 5 N, which acts at 20° to the vertical. Find the tension in the second string and the angle between it and the vertical.

11 A particle of weight 10 N is in equilibrium on a smooth plane, inclined 40° to the horizontal. A horizontal force P acts on the particle. Find P and the normal reaction between the plane and the particle.

12 A particle of weight 10 N is suspended from a fixed point by a light inextensible string. A horizontal force of 5 N also acts on the particle. Find the tension in the string and the angle between the string and the vertical.

13 Three forces act upon a particle, which is in equilibrium. If the magnitudes of the forces are 4 N, 5 N, and 6 N, find the angles between the forces.

14 A and B are particles connected by a light, inextensible string which passes over a smooth fixed pulley attached to a corner of a smooth fixed plane, inclined at 37° to the horizontal. Particle B hangs freely. If the mass of A is 3 kg, and the system is in equilibrium, find the mass of B.

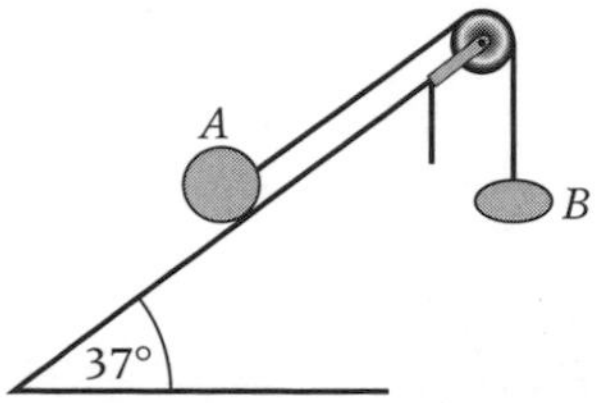

15 An object of mass 3 kg is suspended by two light, incxtensible strings. The strings make angles of 30° and 40° and the horizontal, as shown in the diagram.

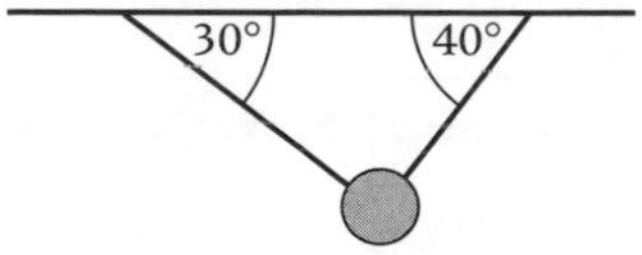

Find the magnitude of the tension in each string.

16 An object, of mass 40 kg, is supported in equilibrium by 4 cables. The forces, in Newtons, exerted by three of the cables, $\mathbf{F}_1$, $\mathbf{F}_2$ and $\mathbf{F}_3$ are given in terms of the unit vectors **i**, **j** and **k** as $\mathbf{F}_1 = 80\mathbf{i} + 20\mathbf{j} + 100\mathbf{k}$, $\mathbf{F}_2 = 60\mathbf{i} - 40\mathbf{j} + 80\mathbf{k}$ and $\mathbf{F}_3 = -50\mathbf{i} - 100\mathbf{j} + 80\mathbf{k}$. The unit vectors **i** and **j** are perpendicular and horizontal and the unit vector **k** is vertically upwards.

Find $\mathbf{F}_4$, the force exerted by the fourth cable, in terms of **i**, **j** and **k**. Also find its magnitude to the nearest Newton. [A]

17 Four boys are playing a "tug of war" game, each pulling horizontally on a rope attached to a light ring. Boy A pulls with a force of $(92\mathbf{i} - 33\mathbf{j})$ N, boy B with force $(66\mathbf{i} + 62\mathbf{j})$ N and by C with force $(-70\mathbf{i} + 99\mathbf{j})$ N, where **i** and **j** are perpendicular unit vectors. Given that the ring is in equilibrium, find the force exerted by boy D, and its magnitude.

4.7 Friction

Consider a body of mass m, placed on a rough horizontal plane. Suppose a horizontal force of magnitude P is applied. If the body remains in equilibrium, then the frictional force, F, will be such that $F = P$.

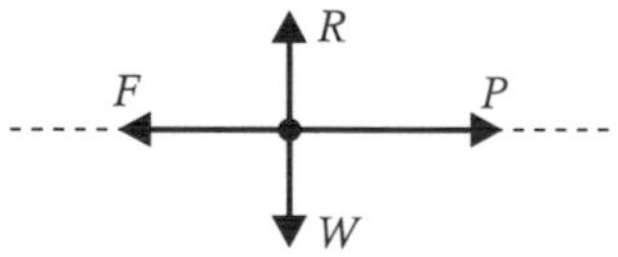

If P is steadily increased and the body remains at rest, then F increases also, so that $F = P$ remains true. However F can only increase up to a certain limit, F_{max}.

> The force between the surfaces in contact and the type of surface determine how large the frictional force can become.

Racing cars make use of this principle. The downwards force on the vehicle is increased by incorporating an aerofoil into the design. This improves the grip of the tyres on the road surface, so that drivers can take corners at greater speed without skidding.

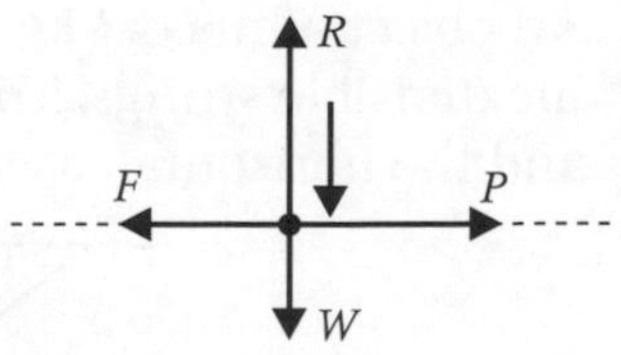

The vertical force between the surfaces is the normal reaction R. If a weight is placed on top of a body so that the normal reaction is doubled, it will be found that F_{max} will also double. That is, F_{max} is proportional to R (the normal reaction).

> $$F_{max} = \mu R$$
>
> where μ is a constant, which depends only on the roughness of the surface, is known as the coefficient of friction.

If P is increased even further then slipping will occur. The frictional force cannot increase further and experiments show that the frictional force remains constant at its maximum throughout the motion.

N.B. (i) When the frictional force does equal its maximum it is often said to be **limiting**. (ii) The special case $\mu = 0$ means $F_{max} = 0$, which corresponds to a smooth plane.

Horizontal planes

Worked example 12

A particle of mass 10 kg is placed on a rough horizontal plane. A horizontal force P acts on the particle. P is increased until the particle is on the point of sliding, which occurs when $P = 10$ N. Find the coefficient of friction between the particle and the plane.

Solution

The diagram shows the forces acting on the particle.

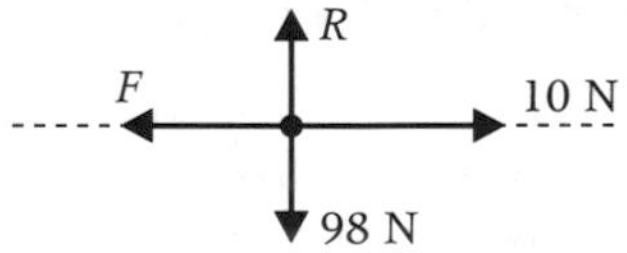

Resolving vertically: $R = 98$ N

Resolving horizontally: $F = 10$ N

If motion is just about to occur, friction is limiting, $F = F_{max}$ so $F = \mu R$.

$$10 = 98\mu$$

$$\mu = \frac{5}{49}.$$

Worked example 13

In the following situations a body of mass 10 kg is placed on a rough horizontal plane. If $\mu = \frac{1}{2}$ in each case determine whether motion will occur.

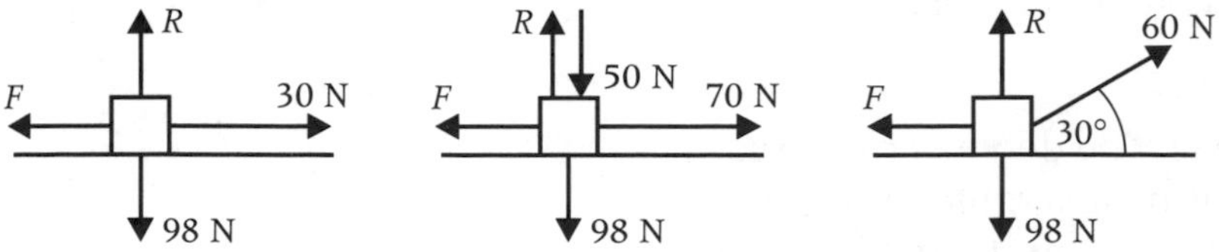

(a) (b) (c)

Solution

(a) Resolving vertically: $R = 98$ N

$F_{max} = 0.5 \times 98 = 49$ N

For equilibrium $F = 30 < 49$, so no motion will occur.

(b) Resolving vertically: $R = 98 + 50 = 148$ N

$F_{max} = 0.5 \times 148 = 74$ N

For equilibrium $F = 70 < 74$, so again no motion will occur.

(c) Resolving vertically: $R + 60 \sin 30° = 98$

$R = 68$ N

$F_{max} = 0.5 \times 68 = 34$ N

For equilibrium $F = 60 \cos 30° = 52 > 34$, which is not possible, so sliding will occur.

Use of the inequality $F \leqslant \mu R$

In all possible cases of a particle in equilibrium on a rough plane, the frictional force has a limiting value so the following inequality is always true:

$$F \leqslant \mu R$$

This inequality itself can be used to good effect.

Worked example 14

A particle, of mass 10 kg, is at rest on a rough horizontal plane. A force, of magnitude of P N, is applied in the direction shown. If $\mu = \frac{1}{2}$, what is the greatest possible value of the magnitude of P such that motion does not occur?

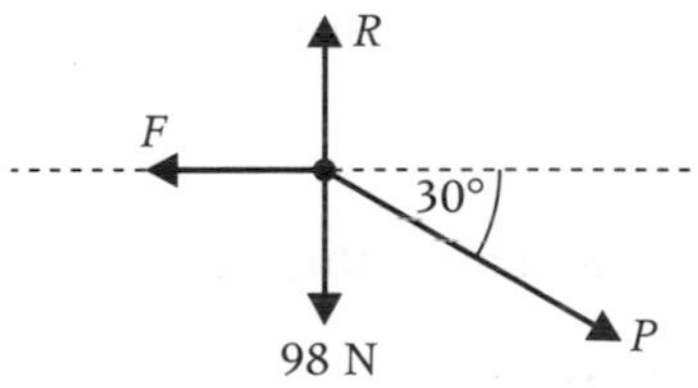

Solution

Resolving horizontally: $F = P \cos 30°$.

Resolving vertically: $R = 98 + P \sin 30°$.

If we substitute for F and R in "$F \leqslant \mu R$"

$$P \cos 30° \leqslant \tfrac{1}{2}(98 + P \sin 30°)$$

Rearranging gives:

$$P(\cos 30° - \tfrac{1}{2} \sin 30°) \leqslant 49$$

$$P \leqslant \frac{49}{\cos 30° - \frac{1}{2} \sin 30°}$$

$$P \leqslant 79.5 \text{ N}$$

So the greatest value of P is 79.5 N (to three significant figures)

Worked example 15

A particle of weight W is at rest on a rough horizontal surface. A force of magnitude W is applied to the particle as shown. Show that the least value of μ is $1 + \sqrt{2}$.

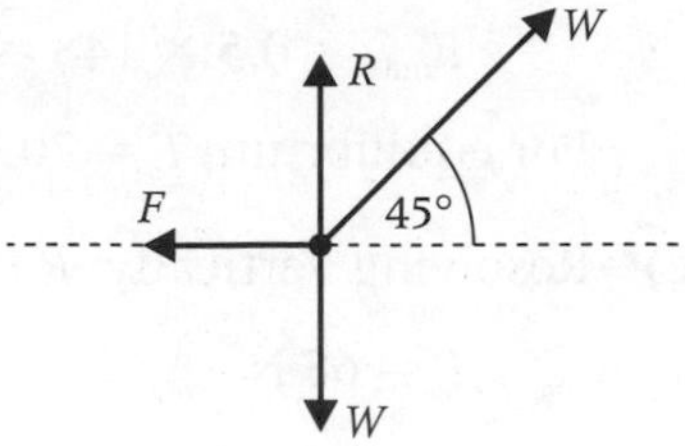

Solution

The diagram shows the forces acting on the diagram.

Resolving horizontally: $F = W\cos 45 = \dfrac{\sqrt{2}}{2}W$

Resolving vertically: $R + W\sin 45 = W$

$$R = W - W\sin 45$$

$$= W\left(1 - \frac{\sqrt{2}}{2}\right)$$

Substituting for F and R in "$F \leqslant \mu R$" we get:

$$\frac{\sqrt{2}}{2}\mathrm{W} \leqslant \mu\mathrm{W}\left(1 - \frac{\sqrt{2}}{2}\right)$$

$$\frac{\sqrt{2}}{2\left(1 - \frac{\sqrt{2}}{2}\right)} \leqslant \mu$$

Rationalising the denominator gives:

$$\frac{\sqrt{2}(2 + \sqrt{2})}{(2 - \sqrt{2})(2 + \sqrt{2})} \leqslant \mu$$

$$\frac{2\sqrt{2} + 2}{2} \leqslant \mu$$

$$1 + \sqrt{2} \leqslant \mu$$

So the least value of μ which satisfies this inequality is $1 + \sqrt{2}$.

EXERCISE 4F

1 A horizontal force of magnitude 5 N is applied to a body of mass 20 kg which is at rest on a rough horizontal plane. Find the coefficient of friction, given that friction is limiting in this position.

2 In the following situations a particle of mass 4 kg is placed on a rough horizontal plane. If $\mu = 5/7$ determine whether motion will occur.

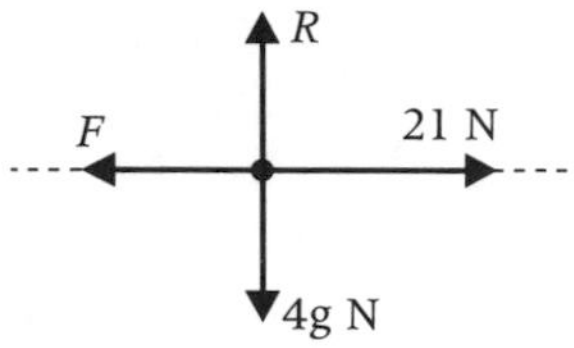

(a)

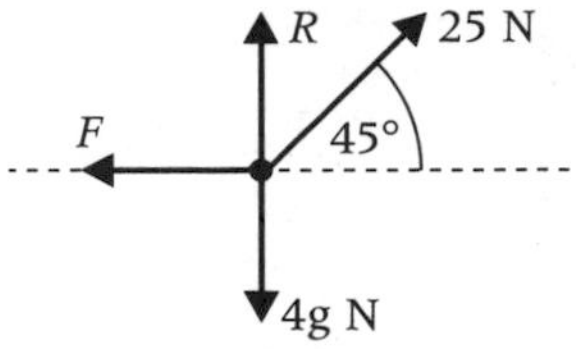

(b)

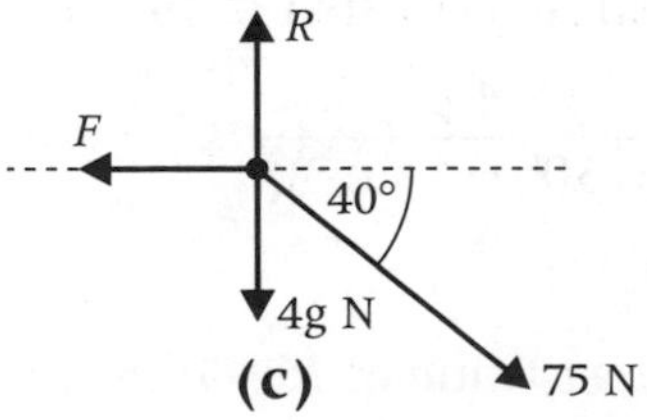

(c)

3 A particle of weight 10 N is at rest on a rough horizontal plane. The particle is pulled by a light, inextensible string, inclined at an angle of 20° to the plane. If the tension in the string is 5 N, find the least value of the coefficient of friction correct to 3 significant figures.

4 The diagram shows a particle of mass of 6 kg at rest on a rough horizontal plane, subject to an external force, of magnitude P N. What is the greatest value of P, if $\mu = 2/3$.

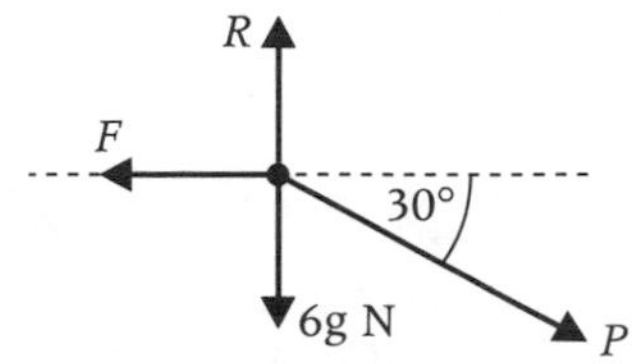

5 Horizontal forces, each of magnitude 100 N act in the direction of the bearings 050°, 170°, and 260°. Find the resultant of these forces. If these forces act on a particle, of mass 10 kg, which is in equilibrium on a rough horizontal plane, find the magnitude of frictional force, which acts on the particle. Find also the least value of the coefficient of friction.

6 Two particles A and B are attached by a light, inextensible string passing over a smooth fixed pulley, as shown, and A has twice the mass of B. Find the smallest value of μ, the coefficient of friction, for equilibrium to be maintained.

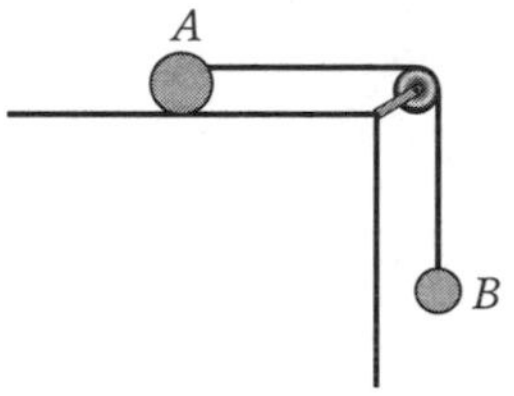

7 The coefficient of friction between a sledge and a snowy surface is 0.2. The combined mass of a child and the sledge is 45 kg. What is the least horizontal force necessary to pull the sledge along the horizontal surface at a constant speed?

8 A sledge of mass 12 kg is on level ground.

(a) A horizontal force of 10 N will keep the sledge moving at a constant speed. Find the value of the coefficient of friction.

(b) A girl of mass 25 kg sits on the sledge. Find the least horizontal force required to keep the sledge moving at a constant speed.

9 A small ring, of weight w N, is threaded on a horizontal curtain rail. A light, inextensible string is pulling it along the rail. The tension in the string is equal to $2w$ N. Show that the least value of the coefficient of friction is $2 - \sqrt{2}$.

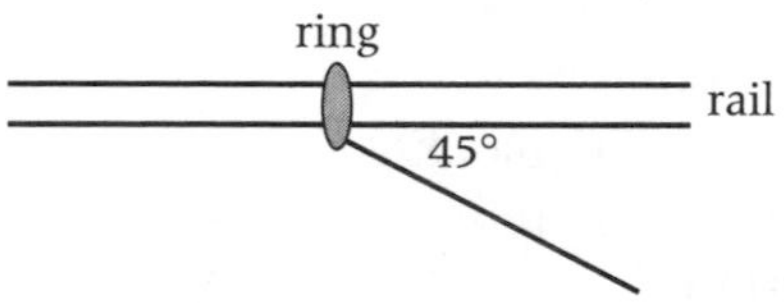

10 The coefficient of friction between a particle, of weight w, and a horizontal plane is μ. The particle is in equilibrium subject to a force of magnitude P N which acts at an angle of θ below the horizontal. Show that:

$$P \leqslant \frac{\mu w}{\cos\theta - \mu\sin\theta}$$

4.8 Friction on inclined planes

Consider a particle, of mass m kg, which is at rest on a rough plane inclined at an angle α to the horizontal.

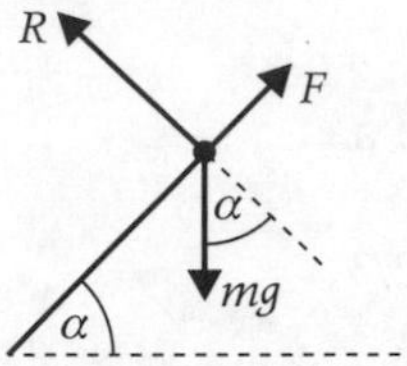

The diagram shows the forces acting on the particle. The friction force must act up the slope to maintain equilibrium.

Resolving parallel to the plane: $F = mg \sin \alpha$

Resolving perpendicular to the plane: $R = mg \cos \alpha$

However

$$F \leqslant \mu R$$

so

$$mg \sin \alpha \leqslant \mu\, mg \cos \alpha$$

hence:

$$\frac{\sin \alpha}{\cos \alpha} \leqslant \mu$$

$$\tan \alpha \leqslant \mu$$

This result tells us that the particle will slip if the angle becomes too steep. The particle will be on the point of slipping when

$$\tan \alpha = \mu$$

or

$$\alpha = \tan^{-1} \mu$$

Worked example 16

A block of mass 3 kg is placed on a rough horizontal table. The table is gradually tilted until the particle begins to slip. The block is on the point of slipping when the table is inclined at an angle of 41° to the horizontal. Find the coefficient of friction.

Solution

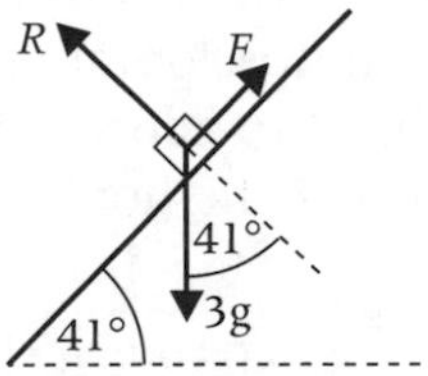

The diagram shows the forces acting on the block.

Resolving parallel to the plane: $F = 3\,g \sin 41° = 19.3$ N

Resolving perpendicular to the plane $R = 3\,g \cos 41° = 22.2$ N

But if the particle is on the point of sliding then friction is limiting so

$$F = \mu R$$
$$19.3 = 22.2\mu$$
$$\mu = 0.869 \text{ (to three significant figures)}$$

Note: We could have used $\mu = \tan^{-1} \alpha$

Suppose this table is now placed at an angle of 50° to the horizontal. The block would not be able to rest in equilibrium unaided. So, suppose a force, of magnitude P, is applied to the block so that it acts up the plane. We will now find the least value of P to maintain equilibrium.

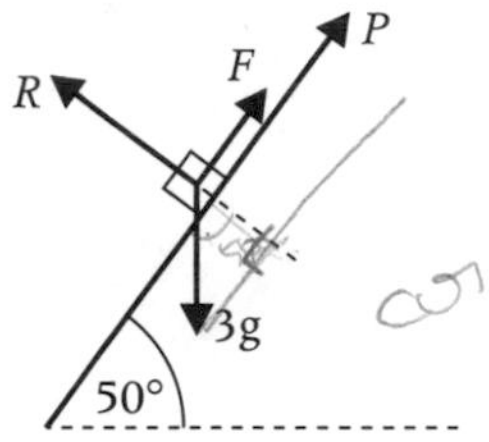

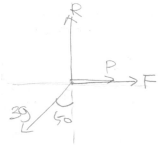

Resolving parallel to plane: $F + P = 3\,g \sin 50°$

$F = 3\,g \sin 50° - P$

Resolving perpendicular to plane: $R = 3\,g \cos 50°$

Here we can use $F \leqslant \mu R$, and substitute for F and R.

$$3\,g \sin 50° - P \leqslant \mu\, 3\,g \cos 50°$$
$$P \geqslant 3\,g \sin 50° - \mu\, 3\,g \cos 50°$$
$$P \geqslant 6.09 \text{ N (to three significant figures)}$$

So the **minimum** value of P is 6.09 N if the particle is to remain at rest.

If the force P is too large, however, the particle will be pulled up the plane so we will now find the **greatest** value of P such that the particle will remain in equilibrium.

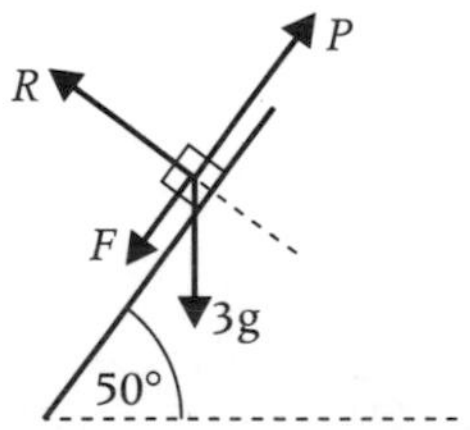

The method used to solve this problem will be exactly the same as that used to find the minimum value of P. There is one fundamental difference, however, which can be seen in the diagram; the frictional force now acts downhill. If we are considering pulling the particle uphill, then friction will oppose the motion and so act downhill.

Resolving parallel to plane: $P = F + 3\,g \sin 50°$

Resolving perpendicular to plane: $R = 3\,g \cos 50°$

Substituting into $F \leqslant \mu R$

$$P - 3\,g \sin 50° \leqslant \mu\, 3\,g \cos 50°$$

$$P \leqslant 3\,g \sin 50° + \mu\, 3\,g \cos 50°$$

$$P \leqslant 38.9 \text{ N}$$

Worked example 17

A body of weight W is placed on a rough plane, inclined at 30° to the horizontal, where $\mu = \frac{1}{\sqrt{3}}$. Find the greatest horizontal force that can be applied to the body, if it is to remain at rest.

Solution

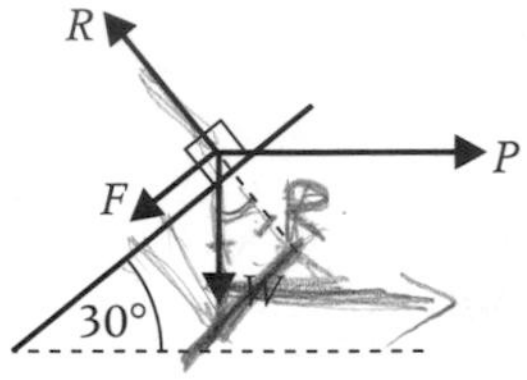

The diagram shows the forces acting on the body.

Resolving parallel to the plane: $P \cos 30° = F + W \sin 30°$

$$F = \frac{\sqrt{3}}{2}P - \frac{1}{2}W$$

Resolving perpendicular to the plane: $R = W \cos 30° + P \sin 30°$

$$R = \frac{\sqrt{3}}{2}W + \frac{1}{2}P$$

Substituting for F and R in $F \leqslant \mu R$ gives:

$$\frac{\sqrt{3}}{2}P - \frac{1}{2}W \leqslant \frac{1}{\sqrt{3}}\left(\frac{1}{2}P + \frac{\sqrt{3}}{2}W\right)$$

$$\frac{\sqrt{3}}{2}P - \frac{1}{2\sqrt{3}}P \leqslant W$$

$$\frac{\sqrt{3}}{2}P - \frac{\sqrt{3}}{6}P \leqslant W$$

$$\frac{\sqrt{3}}{3}P \leqslant W$$

$$P \leqslant \sqrt{3}W$$

So the greatest force which can be applied is $\sqrt{3}W$

EXERCISE 4G

1 In the following situations a particle of mass m kg is placed on a rough plane inclined at an angle α to the horizontal. Determine whether the particle will slide.

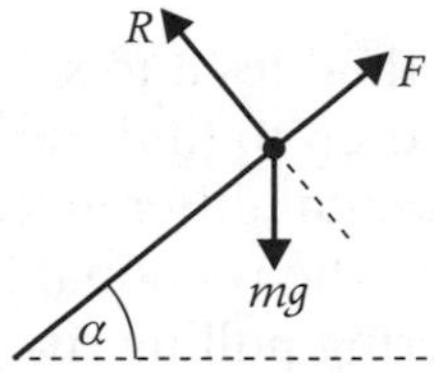

(a) $\alpha = 20°$, $\mu = 0.1$

(b) $\alpha = 30°$, $\mu = 0.75$

(c) $\alpha = 50°$, $\mu = 0.9$

2 The situations below show a particle of mass 10 kg at rest on an inclined plane, where the coefficient of friction is 0.1. Find the magnitude, P, of the applied force, if the particle is on the point of sliding down the plane.

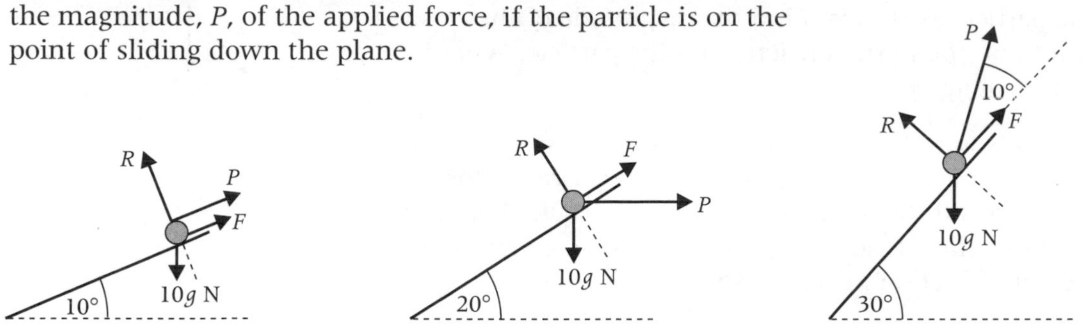

(a) **(b)** **(c)**

3 The situations below show a particle of mass 10 kg at rest on a rough inclined plane, where the coefficient of friction is 0.1. Find the magnitude, P, of the force applied to the particle, if it is on the point of moving up the plane.

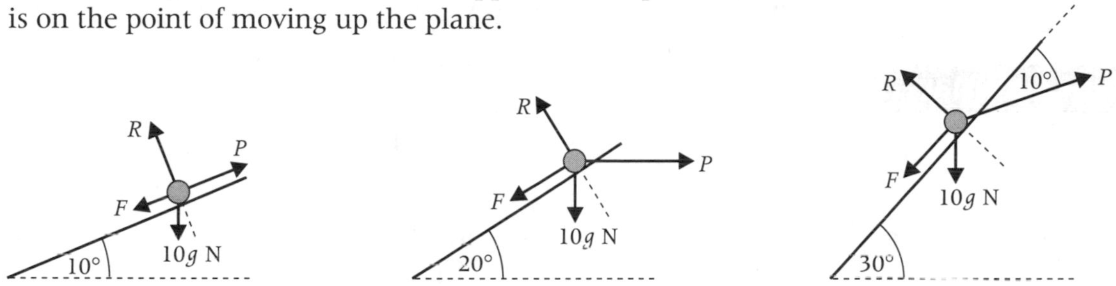

(a) **(b)** **(c)**

4 A particle of mass 10 kg is placed on a rough plane, inclined at 45° to the horizontal. If $\mu = 0.5$ find the least force required to keep the particle in equilibrium, if the force acts upwards, along the line of greatest slope.

5 A particle of mass 6 kg is at rest on a rough plane, of inclination α to the horizontal. Find the greatest horizontal force that can be applied to the particle, if it is to remain in equilibrium, in each of the following cases:

(a) $\alpha = 20°$, $\mu = 0.1$

(b) $\alpha = 30°$, $\mu = 0.9$

(c) $\alpha = 50°$, $\mu = 0.7$.

6 A particle, of mass 12 kg, is at rest on a rough plane, inclined at an angle α to the horizontal. Find the least force, which is parallel to the plane, that must be applied to the particle, if it is to remain in equilibrium, in each of the following cases:

(a) $\alpha = 20°$, $\mu = 0.1$

(b) $\alpha = 30°$, $\mu = 0.5$

(c) $\alpha = 50°$, $\mu = 1.0$.

7 A particle of mass 8 kg is at rest on a rough plane inclined at an angle of 30° to the horizontal. A horizontal force of 20 N acts on the particle as shown. Find the magnitude of the friction force and the normal reaction on the particle. What is the least value of μ?

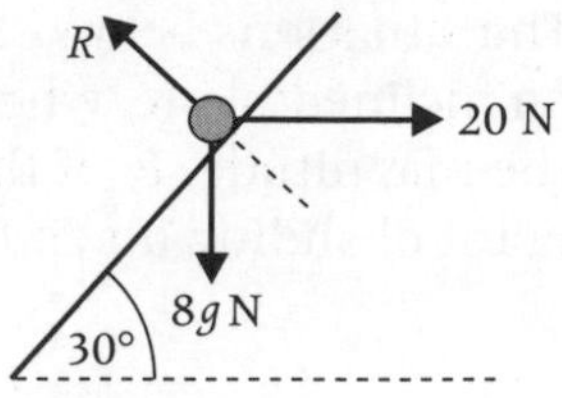

8 The situations below show a particle of mass 4 kg at rest on an inclined plane, subject to a given external force. Draw a force diagram showing all the forces acting on the particle. Find the normal reaction, the friction force, and the range of values of the coefficient of friction in each case.

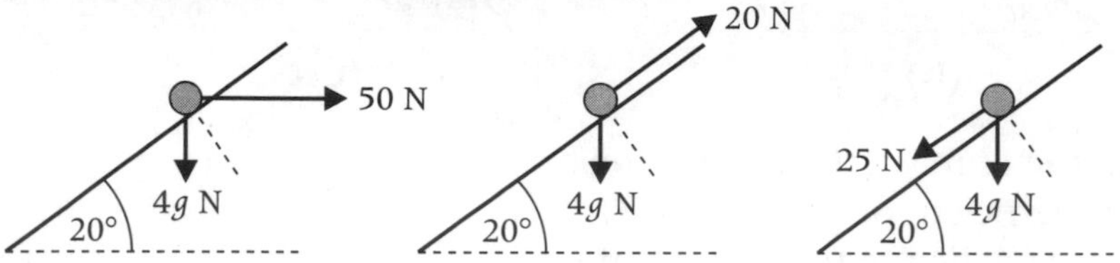

(a) **(b)** **(c)**

Key point summary

Formulae to learn

$F \leq \mu R$

1	Force is a vector quantity.	*p44*
2	Resultant force is the sum of two or more forces. It forms a triangle of forces with two forces and a quadrilateral with three forces.	*p50*
3	Resultant forces can often be found more quickly using forces in component form.	*p51*
4	A force can be divided into two mutually perpendicular components whose vector sum is equal to the given force.	*p53*
5	A particle is in equilibrium if the resultant force acting on it is zero.	*p56*
6	The frictional force between two surfaces in contact will act in a direction to oppose relative motion.	*p61*
7	The magnitude of the frictional force will never be greater than what is necessary to prevent motion.	*p61*
8	The maximum value of the frictional force is given by: $F_{\text{max}} = \mu R$.	*p62*
9	If motion takes place then the magnitude of friction force will be μR.	*p62*

Test yourself	What to review
	If your answer is incorrect – review

1 Draw diagrams to show the forces acting on each of the following. You should include air resistance where appropriate, and model each body as a particle. — *Section 4.2*

(a) A golf ball at its maximum height.

(b) A cyclist travelling up a slope at a constant speed.

(c) A child on a swing at her lowest position.

2 The diagram shows three forces and the perpendicular unit vectors **i** and **j**. — *Section 4.3 and 4.4*

(a) Find the resultant of these three forces in terms of the unit vectors **i** and **j**.

(b) Find the magnitude of the resultant of these three forces and draw a diagram to show the direction in which it acts.

(c) When a fourth force acts at the same point the forces are in equilibrium. Find the magnitude of this force and describe the direction in which it acts.

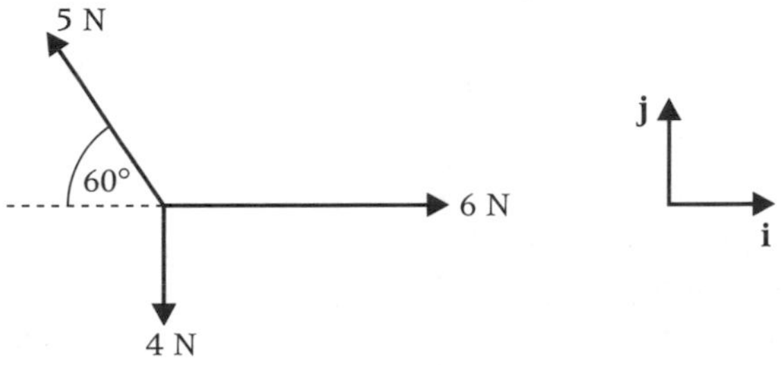

3 The diagram shows an object of mass 50 kg, which is supported by two cables. Find the tension in each of the supporting cables. — *Section 4.5*

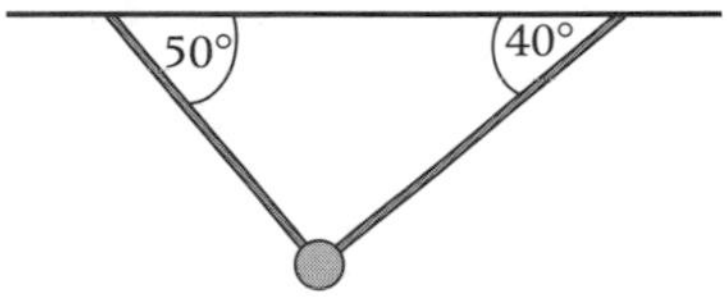

4 The diagram shows a spring fixed to a wall and a block of mass 20 kg, that is at rest on a rough horizontal plane. A string attached to the block passes over a small smooth pulley and is attached to a second block of mass 5 kg. The situation is shown in the diagram. — *Section 4.7*

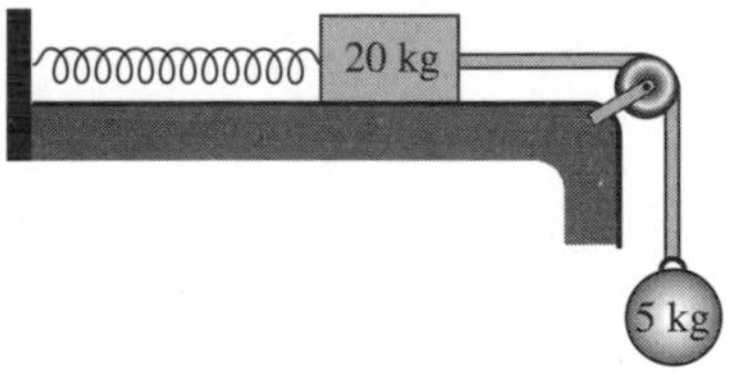

The tension in the spring is 8 N when the block is on the point of sliding towards the pulley.

(a) Find the coefficient of friction between the block and the plane.

(b) Describe what happens to the 20 kg block if the string attached to it is cut. Give reasons to support your answer.

5 A particle, of mass 5 kg, is at rest on a slope inclined at an angle of 48° to the horizontal. A force, of magnitude 10 N, that is directed up the slope acts on the particle. — *Section 4.7*

(a) Find the magnitude of the friction force acting on the particle.

(b) Find an inequality that the coefficient of friction between the particle and the slope must satisfy.

Test yourself ANSWERS

1 (a)

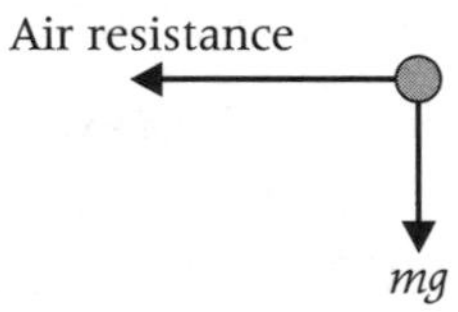

(b)

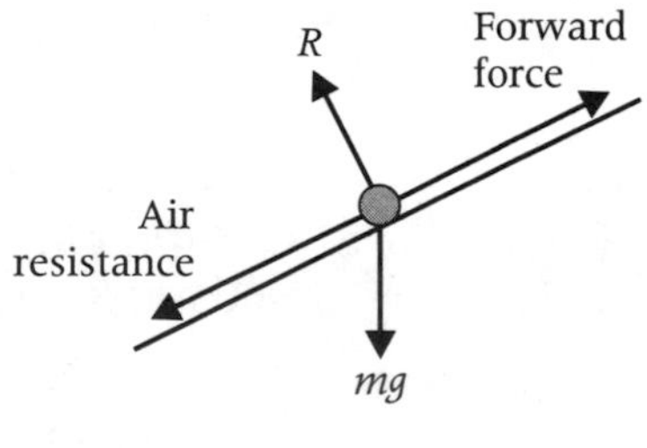

(c)

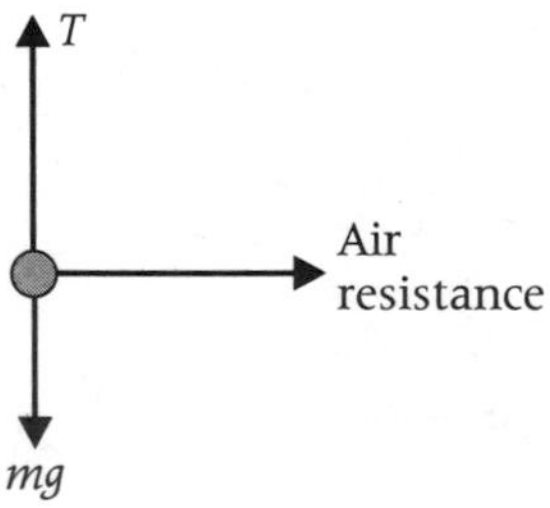

2 (a) $3.5\mathbf{i} + 0.330\mathbf{j}$.

(b) 3.52 N.

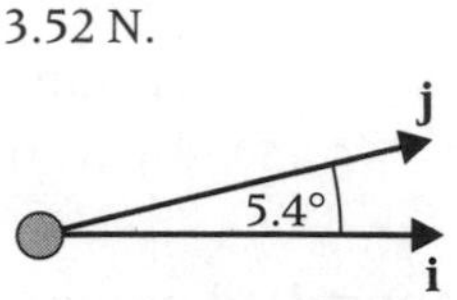

(c) 3.52 N.

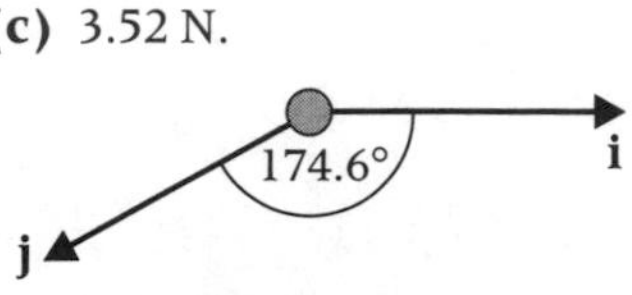

3 315 N, 375 N.

4 (a) 0.209.

(b) Nothing as 8 N is less than $F_{max} = 41$ N.

5 (a) 26.4 N.

(b) $\mu \geqslant 0.806$.

CHAPTER 5

Newton's laws of motion

Learning objective

After studying this chapter you should be able to:

- understand the relationship between force and motion as described by Newton's laws.

5.1 Introduction

In this chapter we will examine particles in motion. The relationship between force and motion is described by Newton's laws. In the situations that we will encounter we will use Newton's laws to enable us to model motion produced by forces. If we were to study the motion of atomic particles it would be found that Newton's laws would not provide an accurate enough model and the theory of relativity should be considered.

5.2 Newton's first law

> *A body will remain in a state of rest or will continue to move in a straight line with a constant velocity unless it is compelled to change that state by the action of a force.*

In the previous chapter we considered the state of equilibrium to be when the resultant force is zero. Newton's first law implies that if a particle moves with a constant velocity then the resultant force will also be zero.

Worked example 1

A particle of mass 3 kg slides down a rough plane at an angle α to the horizontal at constant speed. If $\mu = 0.5$ find the angle α.

Solution

The diagram shows the forces acting on the particle.

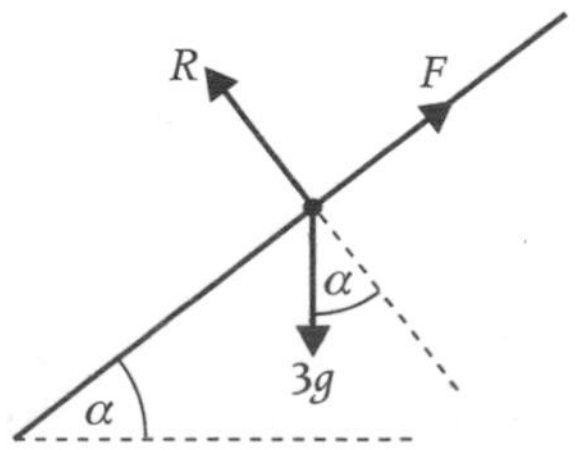

As the particle is moving at a constant speed, the resultant force on the particle is zero.

Resolving horizontally: $mg \sin \alpha = F$

Resolving vertically: $mg \cos \alpha = R$

The laws of friction state that if sliding occurs on a rough surface then the frictional force has reached its maximum and $F = \mu R$, hence

$$F = 0.5R$$

$$mg \sin \alpha = 0.5mg \cos \alpha$$

$$\frac{\sin \alpha}{\cos \alpha} = 0.5$$

$$\tan \alpha = 0.5$$

$$\alpha = 26.6° \text{ (to three significant figures)}$$

Worked example 2

A stone of mass 100 g is dropped in a lake. The stone experiences a resistive force which is proportional to the square of its speed, and reaches a maximum speed of 2 m s^{-1}. What will be the maximum speed of a similar stone of mass 50 g?

Solution

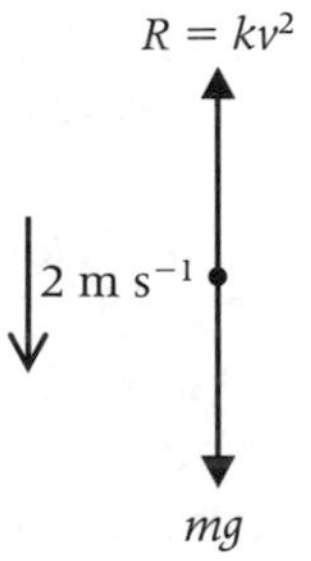

As the stone falls it will increase in speed. The resistive force will therefore increase as well. The maximum speed will be achieved when the resistive force balances the weight of the stone, i.e. $R = mg$.

If R is proportional to v^2 then

$R = kv^2$, where k is a constant.

Hence at maximum speed

$$mg = kv^2$$

$$0.1 \times 9.8 = k \times 2^2$$

$$k = 0.245$$

The resistive force on the similar stone will be a function of its speed and shape, and not influenced by its mass. If we assume it has the same shape as the first stone then we can use $R = 0.245v^2$ again.

When this stone reaches its maximum speed

$$R = mg$$

$$0.05 \times 9.8 = 0.245v^2$$

$$v = 1.41 \text{ m s}^{-1} \text{ (to three significant figures)}$$

EXERCISE 5A

1 The following situations show a body moving with constant velocity in the direction shown, subject to unknown forces of magnitude X and Y. Find X and Y.

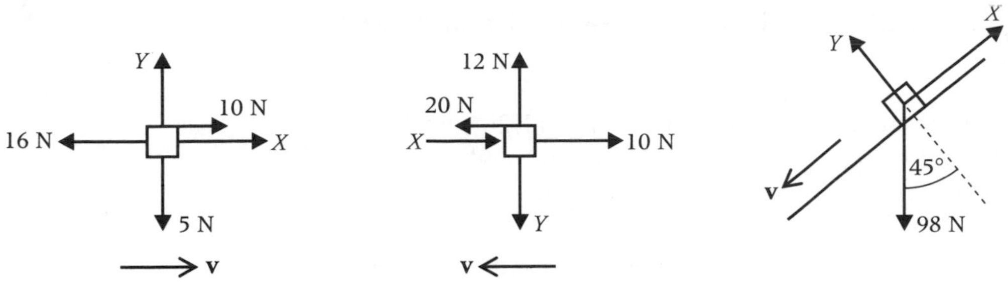

(a) **(b)** **(c)**

2 The following situations show a body moving with constant velocity in the direction shown, subject to an unknown force of magnitude F N acting an angle θ. Find F and θ.

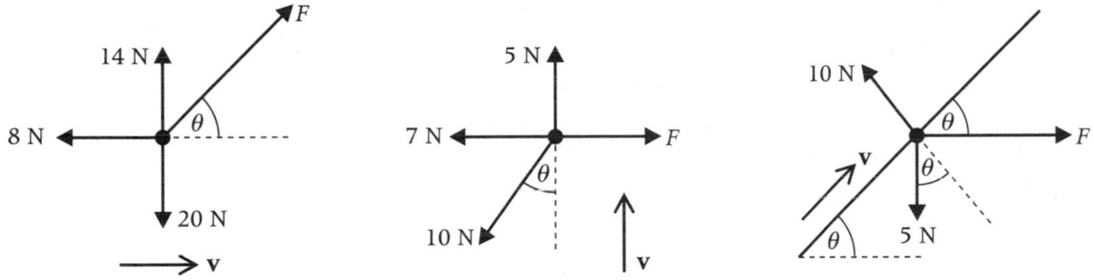

(a) **(b)** **(c)**

3 A lorry travels along a horizontal road with constant velocity. If the force which the engine exerts is 1350 N, what is the magnitude of the resistive force on the lorry?

4 A small object is being pulled across a horizontal surface at a constant velocity by a force of 12 N acting parallel to the surface. If the mass of the object is 5 kg, determine the coefficient of friction between the object and the surface.

5 A sledge of mass 16 kg is being pulled up the side of a hill of inclination 25°, at a constant velocity. The coefficient of friction between the sledge and the hill is 0.4, and the rope pulling the sledge exerts a force of magnitude T N at 15° to the hill.

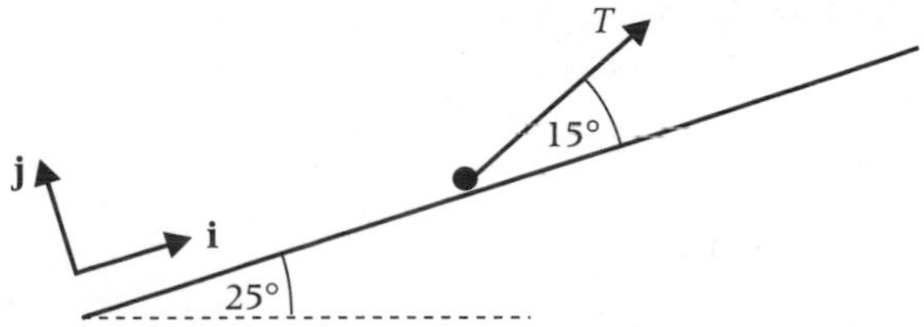

Taking **i** and **j** as unit vectors parallel and perpendicular to the plane, as shown, and modelling the sledge as a particle, write each force as a vector and find:

(a) the magnitude of the tension in the rope, and

(b) the magnitude of the normal reaction between the hill and the sledge.

6 A car of mass 1 tonne travels up an incline of $\sin^{-1}(1/20)$, with constant velocity. If the engine exerts a force of 700 N, what is the magnitude of the resistive force on the car?

7 A lorry of mass 20 tonnes can climb an incline of $\sin^{-1}(1/20)$ at a steady speed of 10 m s^{-1}, when the force produced by the engine is 10.2 kN. Assume that the air resistance is proportional to the square of the speed and ignoring other forms of resistance, find the maximum speed of the lorry when freewheeling down the same hill.

8 A lorry when fully laden weighs 40 tonnes. Its maximum speed when freewheeling down an incline of $\sin^{-1}(1/10)$ is 40 m s^{-1}, subject to air resistance which is proportional to the square of the speed of the lorry. When empty the lorry weighs 8 tonnes. What would be the maximum speed of the empty lorry when freewheeling down the same hill?

9 A skier of mass 80 kg can achieve a maximum speed of 35 m s^{-1} down an incline 30° to the horizontal, subject to: air resistance which is proportional to her speed; and friction where the coefficient of friction between skier and the slope is 0.1. What will be the maximum speed of the skier down an incline of 45° to the horizontal?

5.3 Newton's second law

Newton's second law describes the relationship between the change in motion of a body and the resultant force acting on the body. Provided the mass of the object concerned does not change, as it might if it was a space rocket, then the law can be stated mathematically as:

> $\boldsymbol{F} = m\boldsymbol{a}$
>
> where $\boldsymbol{F}$ is the resultant force on the body, which has mass m and $\boldsymbol{a}$ is its acceleration.

Note that:

(i) $\boldsymbol{F}$ and $\boldsymbol{a}$ are vector quantities. So the acceleration is in the direction of the resultant force.

(ii) The unit of mass is the kg. We measure acceleration in terms of $\mathrm{m\,s^{-2}}$. A force of one newton (N) is defined as that force necessary to give a mass of 1 kg an acceleration of $1\,\mathrm{m\,s^{-2}}$.

Worked example 3

Find the acceleration produced by a force of 20 N on a mass of 4 kg.

Solution

Using $F = ma$ with $F = 20$ and $m = 4$ gives:

$$20 = 4a$$

$$a = 5\,\mathrm{m\,s^{-2}}$$

Worked example 4

Find the acceleration produced by the forces $(\mathbf{i} + 3\mathbf{j})$ N, $(3\mathbf{i} + \mathbf{j})$ N and $(4\mathbf{i} + 2\mathbf{j})$ N which act on a mass of 2 kg.

Solution

The resultant force is:

$$\mathbf{i} + 3\mathbf{j} + 3\mathbf{i} + \mathbf{j} + 4\mathbf{i} + 2\mathbf{j} = 8\mathbf{i} + 6\mathbf{j}.$$

Using $F = ma$

$$2a = 8\mathbf{i} + 6\mathbf{j}$$

$$a = 4\mathbf{i} + 3\mathbf{j}$$

Note: the magnitude of the acceleration is $\sqrt{4^2 + 3^2} = 5\,\mathrm{m\,s^{-2}}$

Worked example 5

A particle of mass 3 kg is pulled across a rough horizontal plane, by a light inextensible string inclined at 30° to the horizontal. The tension in the string is 40 N. The coefficient of friction between the particle and the plane is 0.5. Find the acceleration of the particle.

Solution

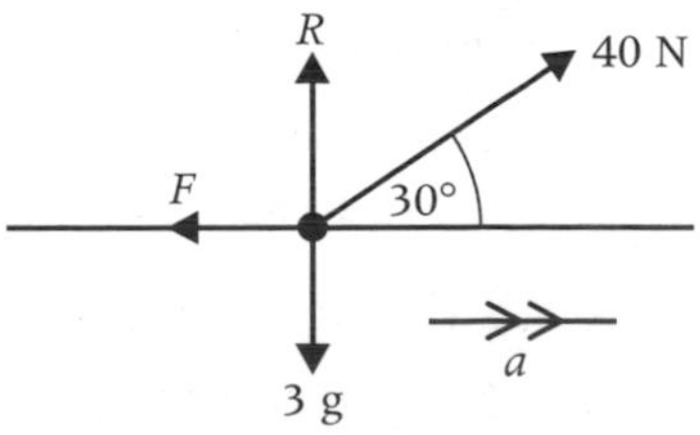

The diagram shows the forces acting and the acceleration of the particle.

The particle accelerates along the plane so the resultant force must also be parallel to the plane. So the sum of the vertical components of the forces must be zero.

Resolving vertically:

$$R + 40 \sin 30° - 3\,g = 0$$

$$R = 9.4\text{ N}$$

The frictional force will be at its maximum, i.e.

$$F = \mu R = 0.5 \times 9.4$$

$$F = 4.7\text{ N}$$

Using $F = ma$ parallel to the plane:

$$40 \cos 30° - 4.7 = 3a$$

$$a = 9.98\text{ m s}^{-2}$$

Worked example 6

A stone is projected with speed 10 m s^{-1} down the line of greatest slope of a plane inclined at 30° to the horizontal. The coefficient of friction between the slope and the stone is 0.8. Find the distance travelled by the stone before coming to rest.

Solution

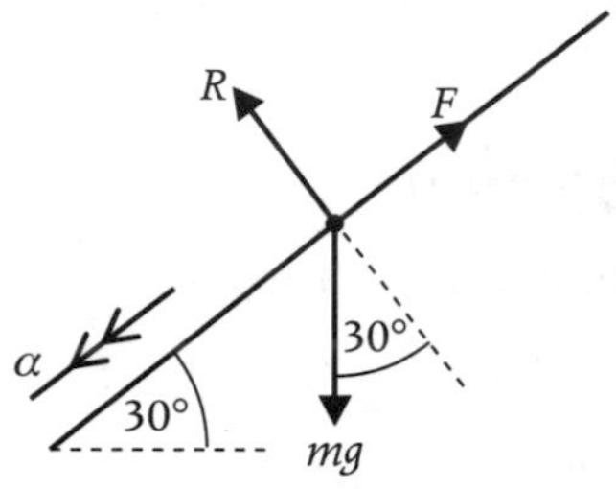

The diagram shows the forces acting and the acceleration of the particle. The first step is to find the acceleration of the stone.

The components of the forces perpendicular to the plane must be in equilibrium. So resolving perpendicular to the plane gives

$$R = mg \cos 30°.$$

Friction is limiting, i.e.

$$F = \mu R$$

$$= 0.8\,mg \cos 30°$$

Using Newton's second law parallel to the plane gives:

$$mg \sin 30° - F = ma$$

$$mg \sin 30° - 0.8\,mg \cos 30° = ma$$

The mass cancels throughout this equation and we get:

$$a = -1.89\text{ m s}^{-2} \text{ (to three significant figures)}$$

Throughout this motion all the forces are constant so the acceleration will also be constant. If the acceleration is constant we can use the constant acceleration formulae.

Using

$$v^2 = u^2 + 2as$$

gives

$$0^2 = 10^2 + 2 \times (-1.89)d$$

where d is the distance the stone slides. This gives

$$d = 26.5\text{ m}$$

EXERCISE 5B

1 A single force of magnitude 8 N acts on a particle of mass 16 kg. Find the acceleration produced.

2 The acceleration of a particle of mass 2 kg is 3 m s^{-2}. Find the magnitude of the resultant force on the particle.

3 Forces of $(\mathbf{i} + \mathbf{j})$ N, $(3\mathbf{i} - 2\mathbf{j})$ N and $(\mathbf{i} + 3\mathbf{j})$ N act on a particle of mass 2 kg. Find the acceleration of the particle in the form $a\mathbf{i} + b\mathbf{j}$.

4 A crane lifts a load of mass 2 tonnes vertically from rest. The tension in the cable is 21 kN. Find the acceleration of the load.

5 A package with mass 300 kg is lifted vertically upwards. Find the tension in the cable which lifts the package, when the package:

(a) accelerates upwards at 0.1 m s^{-2}

(b) accelerates downwards at 0.2 m s^{-2}

(c) travels upwards with a retardation of 0.1 m s^{-2}.

6 Find the unknown accelerations, forces and angles in the following situations.

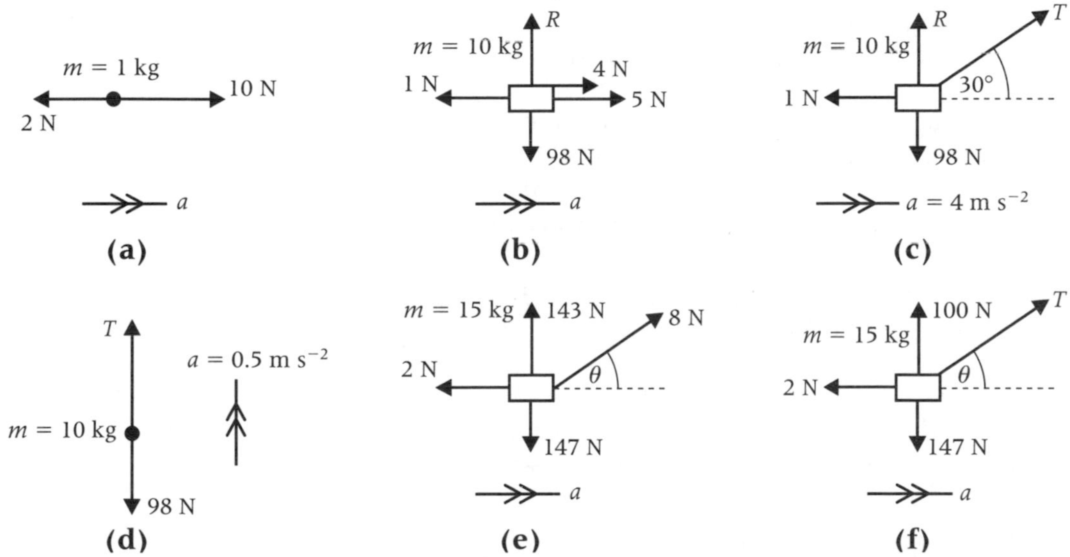

7 Two forces of magnitude 5 N and 6 N act on a particle of mass 2 kg. If the angle between the forces is 60° find the magnitude of the acceleration.

8 Two forces of magnitude 5 N and 6 N act on a particle of mass 2 kg. Find the angle between the forces if the acceleration is 4 m s^{-2}.

9 A car of mass 1 tonne travels along a horizontal road and brakes from $50\,\text{m s}^{-1}$ to rest in a distance of 300 m. Find the braking force on the car.

10 A particle of mass m kg slides down a smooth plane, inclined at an angle of 30° to the horizontal. Find the acceleration of the particle down the plane.

11 A particle of mass 6 kg starts from rest and accelerates uniformly. The resultant force on the particle has magnitude 15 N. Find the time taken to reach a speed of $10\,\text{m s}^{-1}$.

12 A particle of mass 20 kg is pulled across a rough horizontal plane by a light inextensible string, inclined at 30° to the horizontal. If the tension in the string is 50 N and the acceleration produced is $0.5\,\text{m s}^{-2}$ find the frictional force on the particle and the coefficient of friction.

13 A particle of mass m kg slides down a smooth inclined plane with an acceleration of $2\,\text{m s}^{-2}$. Find the inclination of the plane to the horizontal.

14 A crane lifts a load of 5 tonnes through a vertical height of h metres. The load starts from rest and accelerates uniformly for 2 seconds to a speed of $0.5\,\text{m s}^{-1}$. It then travels at constant speed for 20 seconds, after which it is brought to rest in 4 seconds. Find the tension in the cable during each part of the motion, and the distance travelled by the load.

15 A particle of mass 2 kg starts from rest, and accelerates uniformly, subject to a single force, of magnitude P N, until it reaches a speed $30\,\text{m s}^{-1}$. It then travels with constant velocity for 10 seconds, after which time a single force of magnitude $2P$ N, acting in the opposite direction to its velocity brings the particle to rest. If the total distance travelled in the motion is 435 m find P.

16 A particle of mass 500 grams starts from rest at the point A which has position $(4\mathbf{i} + 2\mathbf{j})$ m. The resultant force on the particle is $(\mathbf{i} + 2\mathbf{j})$ N. Find the position vector of the particle after 3 seconds.

17 Two tugs are towing a large oil tanker into harbour. Tug A's engines can produce a pulling force of 80 kN while tug B's engines can produce 65 kN of force.

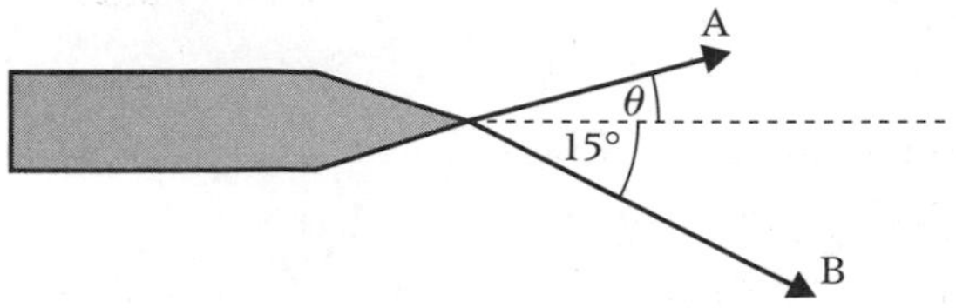

(a) Calculate the angle θ necessary for the tanker to move directly forwards.

(b) Given that there is a resistance to the motion of the tanker of 25 kN directly opposing motion, find the magnitude of the resultant force on the tanker to the nearest 100 N. Find also the acceleration of the tanker if it has a mass of 20 000 tonnes.

18 A sledge of mass 30 kg is accelerating down a hill while a boy is trying to prevent it from sliding by pulling on a rope attached to the sledge with a force of 40 N. The hill has inclination 28° and the rope is inclined to the hill at 10°. The coefficient of friction between the sledge and the hill is 0.35. Find:

(a) the magnitude of the normal contact force on the sledge

(b) the resultant force on the sledge acting down the hill

(c) the magnitude of the sledge's acceleration.

19 As a lift moves upwards from rest it accelerates at 0.8 m s^{-2} for 2 seconds, then travels 4 m at constant speed and finally slows down, with a constant deceleration, stopping in 3 seconds. The mass of the lift and its occupants is 400 kg.

(a) Find the total distance travelled by the lift and the total time taken.

(b) The lift hangs from a single light, inextensible cable. During which stage of the motion of the lift is the tension in the cable greatest? Find the magnitude of this tension during this stage of the motion. [A]

20 A tree trunk, of mass 250 kg, is pulled up a slope by a chain attached to a tractor. The chain is at an angle of 10° to the slope. The slope itself is at 8° to the horizontal. The tree trunk initially accelerates at 0.2 m s^{-2}. A friction force, of magnitude 2000 N, acts on the tree trunk.

(a) Model the tree trunk as a particle. Draw and label a diagram to show the forces acting on it.

(b) Find the initial tension in the chain.

(c) Explain why the tension in the chain will probably decrease. [A]

21 An amateur inventor is trying to design a simple device that will warn car drivers if their acceleration or deceleration is excessive. A prototype is shown in the diagram, and consists of two strings and a small sphere of mass m attached to the roof of a car at the points A and C.

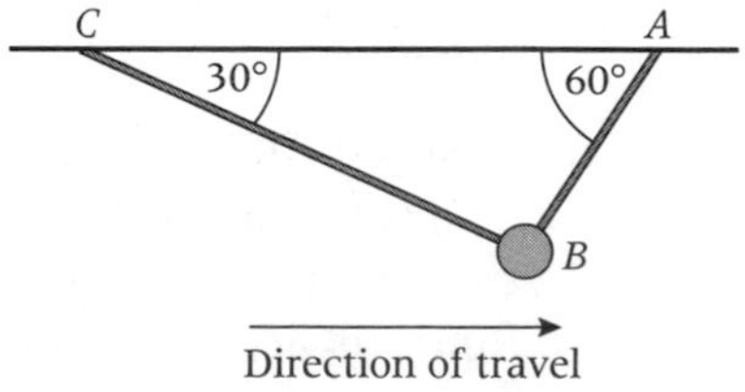

Assume that the car travels in a straight line and that the plane ABC is vertical and parallel to the direction of travel.

Also assume the line AC is horizontal. The acceleration of the car is a.

(a) Show that the tension in the string BC is

$$\frac{m(g - a\sqrt{3})}{2}$$

and find the tension in the other string.

(b) Find the values of a for which each string becomes slack. Is it likely that either string will become slack in normal use? Give reasons for your answer.

(c) Would using a lighter mass reduce the magnitude of the acceleration required for the strings to become slack? Give a reason for your answer. [A]

22 A train is travelling at a constant speed of $40\ \text{m s}^{-1}$, when the driver sees a warning light. Over the next 1000 m the speed of the train drops to $20\ \text{m s}^{-1}$. The train travels at this speed for 5 minutes. The speed returns to $40\ \text{m s}^{-1}$ after a further 5 minutes. Assume that the acceleration of the train is constant on each stage of its journey.

(a) Find the total distance travelled by the train, while its speed is less than its normal operating speed of $40\ \text{m s}^{-1}$.

(b) The train would normally have travelled this distance at a constant $40\ \text{m s}^{-1}$. Find the time by which it was delayed.

(c) The train has a mass of 50 tonnes and moves on a straight, horizontal track. Assume that while the train is slowing down, it experiences a constant resistive force of 40 000 N and one other force parallel to the track. Find the magnitude of this force and state the direction in which it acts. [A]

23 A child is sliding at a constant speed of $4\ \text{m s}^{-1}$ down a long slide. The child has a mass of 45 kg. The slide is inclined at an angle of 40° to the horizontal. Assume that a constant friction force, of magnitude 89 N, acts on the child.

(a) Use the data given to explain why air resistance must be taken into account when modelling the motion of the child. Find the magnitude of the air resistance acting on the child, when he is travelling at a constant speed of $4\ \text{m s}^{-1}$.

(b) Assume that the magnitude of the air resistance is proportional to the speed of the child. The next time that he uses the slide he starts from rest and accelerates. Find his acceleration, when he is moving at $1\ \text{m s}^{-1}$. [A]

5.4 Newton's third law

Newton's third law states:

> For every action, there is an equal but opposite reaction.

This law is often misunderstood, but is really very simple. The best way to understand it is to consider some simple examples.

If you stand on a table, your feet push down on the table. In response to this the table exerts an upward force on your feet. This upward force is the equal but opposite reaction to the downward force you exert on the table. These are both normal reaction forces.

The diagrams show these normal reaction forces (R).

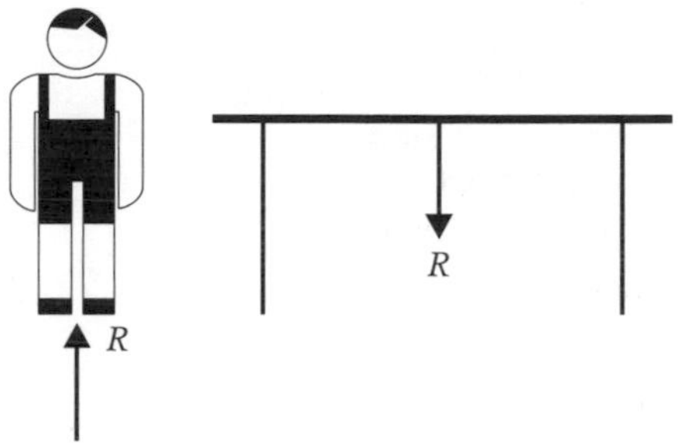

In this example it is important to note that this example does not directly involve the weight of the person (W), although $W = R$.

A second example concerns the orbit of the moon around the earth. The earth exerts a gravitational force on the moon and the moon exerts an equal but opposite force on the earth. The effect of the force acting on the moon is to keep it in orbit around the earth. The effect of the force on the earth is to cause the tides. The important idea, however, is that both planets exert forces of equal size on each other, as shown in the diagram.

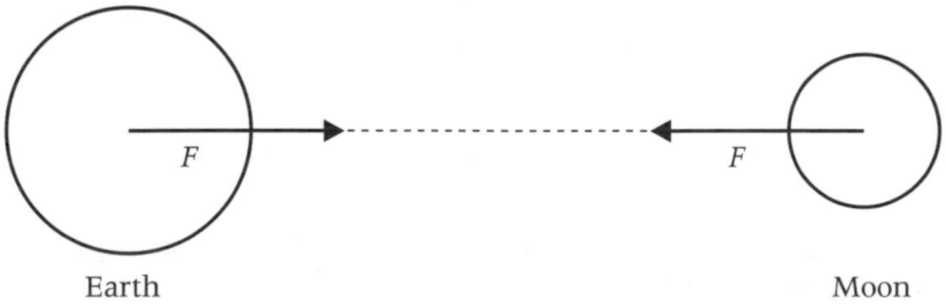

As a final example consider a car towing a caravan. The car exerts a forward force on the caravan, but the reaction to this is the backwards force that the caravan exerts on the car.

EXERCISE 5C

1 Use Newton's third law to explain why you would hurt your hand if you punched a hard object with it.

2 A ladder leans against a smooth wall.

(a) Draw diagrams to show the force that the top of the ladder exerts on the wall and the force that the wall exerts on the top of the ladder.

(b) Use Newton's third law to explain why these forces have equal magnitudes.

3 A child jumps off a table and lands on the ground. Describe how the force that the ground exerts on the child varies. Also describe how the force that the child exerts on the ground varies.

4 A man, of mass 78 kg, stands in a lift of mass 200 kg that is accelerating upwards at 0.5 m s^{-2}. Calculate the magnitudes of the forces that act on the lift. Also draw a diagram to show how they act.

5 Three carriages are coupled to an engine on a set of railway lines. The carriages and engine move forward on horizontal tracks. Draw diagrams to show the forces acting on each of the carriages and the trucks. Clearly show any forces that have the same magnitude.

Key point summary

Formula to learn

$$F = ma$$

1 If a body is at rest or moves with a constant velocity the forces acting on it must be in equilibrium. *p73*

2 When applying Newton's second law, remember that F represents the resultant force. *p76*

3 When two bodies interact, the force exerted by the first body on the second body is equal and opposite to the force exerted by the second body on the first. *p83*

Test yourself	What to review
	If your answer is incorrect – review
1 A helicopter of mass 880 kg is rising vertically at a constant rate. Find the magnitude of the lift force acting on the helicopter. How would your answer change if the helicopter was descending at a constant rate?	*Section 5.2*
2 A child, of mass 30 kg, slides down a slide at a constant speed. Assume that there is no air resistance acting on the child. The slide makes an angle of 40° with the horizontal. Find the magnitude of the friction force on the child and the coefficient of friction.	*Section 5.2*
3 A lift and its passengers have a total mass of 300 kg. Find the tension in the lift cable if: **(a)** it accelerates upwards at 0.2 m s^{-2} **(b)** it accelerates downwards at 0.05 m s^{-2}.	*Section 5.3*
4 A van, of mass 1200 kg, rolls down a slope, inclined at 3° to the horizontal and experiences a resistance force of magnitude 400 N. Find the acceleration of the van.	*Section 5.3*

5

Test yourself ANSWERS

1 8624 N, no change.

2 189 N, 0.839.

3 **(a)** 3000 N **(b)** 2925 N.

4 0.180 m s^{-2}.

CHAPTER 6

Connected particles

Learning objective

After studying this chapter you should be able to:

- solve problems of motion involving connected particles.

6.1 Introduction

In this chapter we will study the motion of bodies that are connected in some way. This will focus mainly on objects that are connected by a light, inextensible string, but we will also look at some other similar examples, for example when a car tows a caravan.

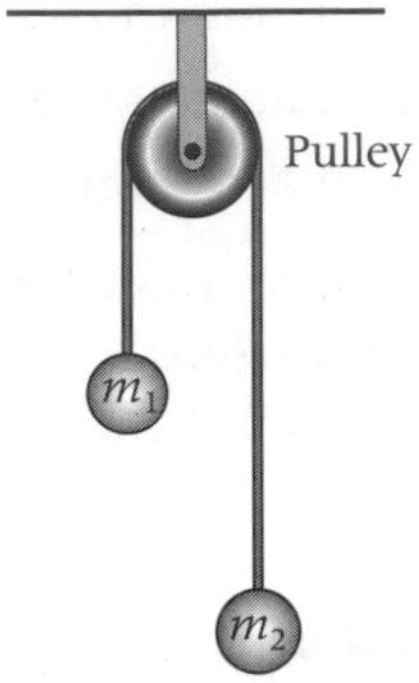

6.2 Motion of two connected particles

When two particles are connected by a light string, it is important to realise that the tension will have the same magnitude throughout the string, although the tension may act in different directions. This understanding of the tension in a string is the key to solving problems that involve connected particles.

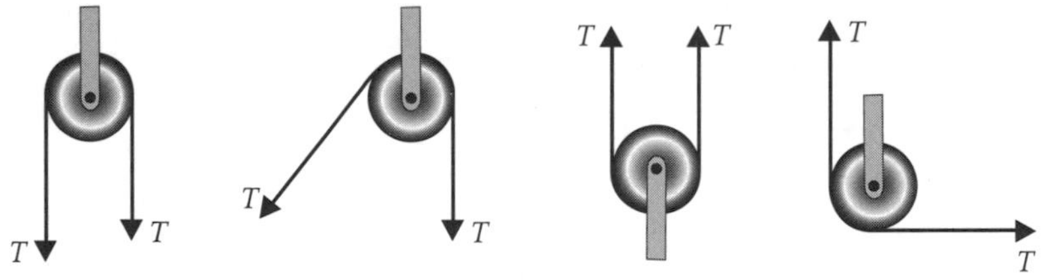

We will only consider cases where the string joining the two objects remains taut, so that both objects have the same acceleration.

When attempting problems with connected particles it is important to consider each particle separately and draw a force diagram for each particle.

Worked example 1

Two particles, of mass 3 kg and 7 kg, are connected by a light, inextensible string. The string passes over smooth pulley, as shown in the diagram.

Find the acceleration of the masses and the tension in the string.

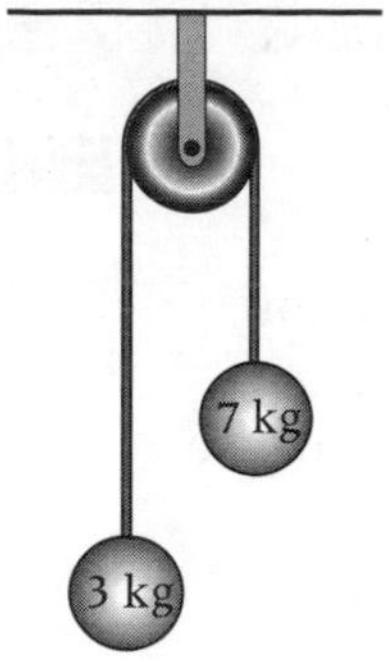

Solution

First consider the 3 kg mass. Two forces act on this mass, its weight downwards and the tension upwards.

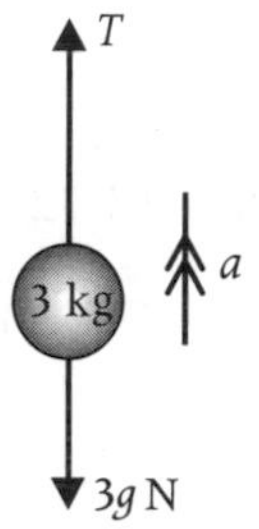

The resultant of these two forces is $T - 3g$.

Applying Newton's second law ($F = ma$) gives the equation below.

$$T - 3g = 3a \qquad (1)$$

where a is the acceleration of the masses.

Now consider the 7 kg mass. Similarly the resultant force on this particle is $7g - T$ and applying Newton's second law gives the equation below.

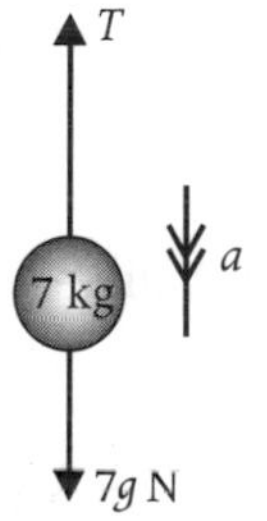

$$7g - T = 7a \qquad (2)$$

These two equations are a pair of simultaneous equations and can now be solved.

From equation (1) $T = 3a + 3g$. This can then be used to substitute for T in equation (2) to give:

$$7g - (3a + 3g) = 7a$$

$$4g = 10a$$

$$a = \frac{4g}{10}$$

$$= 3.92 \text{ m s}^{-2}$$

This acceleration can then be substituted into $T = 3a + 3g$ to give T.

$$T = 3a + 3g$$

$$= 3 \times 3.92 + 3 \times 9.8$$

$$= 41.16 \text{ N}$$

6

Worked example 2

The diagram shows a particle, of mass 7 kg, sliding on a rough horizontal surface and connected by a light, inextensible string to another particle, of mass 5 kg. The coefficient of friction between the surface and the particle is 0.2. Find the acceleration of each particle and the tension in the rope.

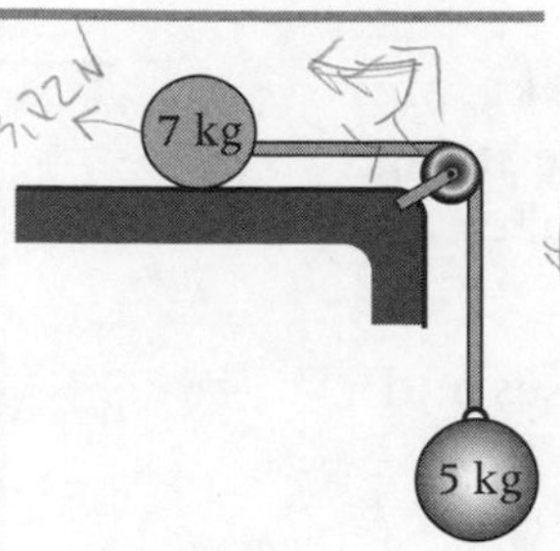

Solution

First consider the particle on the surface. The weight of the particle is 68.6 N, which is balanced by the normal reaction R, so that $R = 68.6$ N.

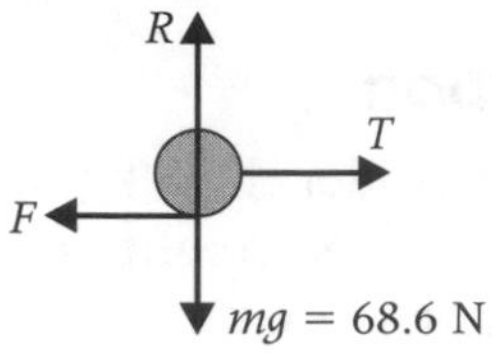

As the particle is sliding the friction force takes its maximum value of μR, so that $F = 0.2 \times 68.6 = 13.72$ N.

So the resultant force, to the right, on this particle is:

$$T - F = T - 13.72$$

Using Newton's second law, $F = ma$, gives:

$$7a = T - 13.72 \qquad (1)$$

Now consider the suspended particle. The resultant downward force on this particle is:

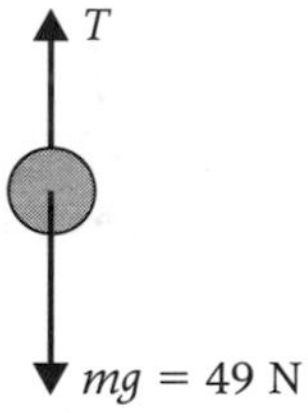

$$49 - T$$

Using Newton's second law, $F = ma$, gives:

$$5a = 49 - T \qquad (2)$$

We now have a pair of simultaneous equations that need to be solved.

From equation (1), $T = 7a + 13.72$, and substituting this into equation (2) gives:

$$5a = 49 - (7a + 13.72)$$

$$12a = 49 - 13.72$$

$$a = \frac{35.38}{12} = 2.94 \text{ m s}^{-2}$$

This can now be used to find T, by substituting for a in the equation:

$$T = 7a + 13.72.$$

$$T = 7 \times 2.94 + 13.72$$

$$= 34.3 \text{ N}$$

Worked example 3

A truck is pulled up a slope by a rope that passes over a pulley and is attached to a large barrel of water, which hangs over the edge of a cliff. The slope is at an angle of 30° to the horizontal and the mass of the truck and its contents is 500 kg. Assume that a constant resistance force of 100 N acts on the truck. The barrel and water have a total mass of 1000 kg.

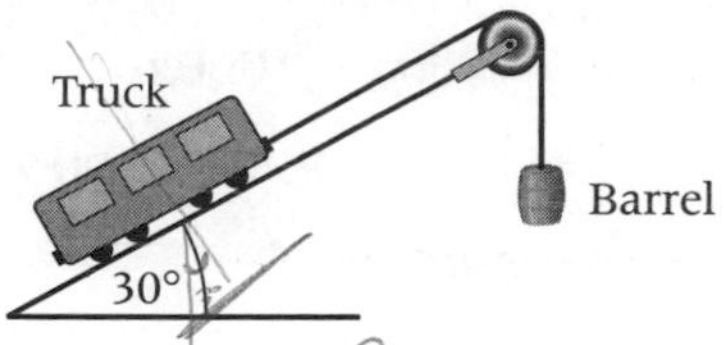

(a) State any assumptions that you need to make.

(b) Find the acceleration of the truck and the tension in the rope.

Solution

(a) The following assumptions should be made:

- the rope is light and inextensible
- there is no friction in the pulley
- the barrel and truck can be modelled as particles
- truck moves up the slope.

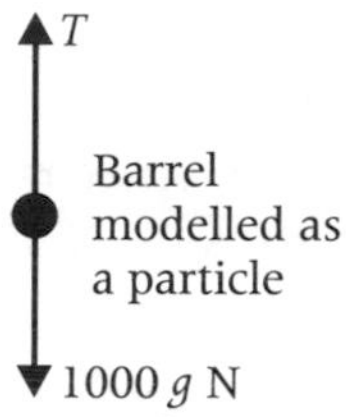

(b) Consider first the forces acting on the barrel.

The resultant force is $1000g - T$, so applying Newton's second law gives the equation below.

$$1000g - T = 1000a \qquad (1)$$

Now consider the forces on the truck. Resolving parallel to the slope gives the resultant force as:

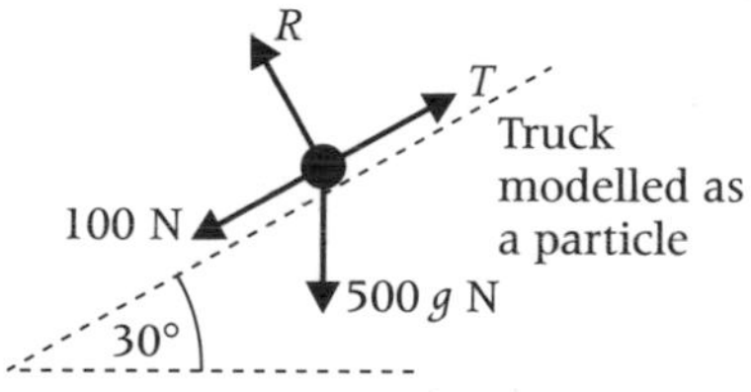

$$T - 100 - 500g\cos 60° = T - 100 - 250g$$

Applying Newton's second law gives:

$$T - 100 - 250g = 500a \qquad (2)$$

We now have a pair of simultaneous equations that can be solved.

From equation (1) $T = 1000g - 1000a$.

This can be substituted into equation (2) to give:

$$T - 100 - 250g = 500a$$

$$1000g - 1000a - 100 - 250g = 500a$$

$$750g - 100 = 1500a$$

$$a = \frac{750g - 100}{1500}$$

$$= \frac{29}{6}$$

$$= 4.83 \text{ m s}^{-2}$$

6

Now the value for a can be substituted into $T = 1000g - 1000a$, to find the tension.

$$T = 1000g - 1000a$$
$$= 1000 \times 9.8 - 1000 \times \frac{29}{6}$$
$$= 4967 \text{ N}$$

Worked example 4

A car of mass 1100 kg pulls a caravan of mass 900 kg. The action of the engine causes a force of magnitude 3000 N to act on the car. Resistance forces of magnitude 300 N and 450 N act on the car and caravan, respectively.

Find the acceleration of the car and caravan and the force that the car exerts on the caravan.

Solution

First consider the car and caravan as a single particle. The resultant force is $3000 - 300 - 450 = 2250$.

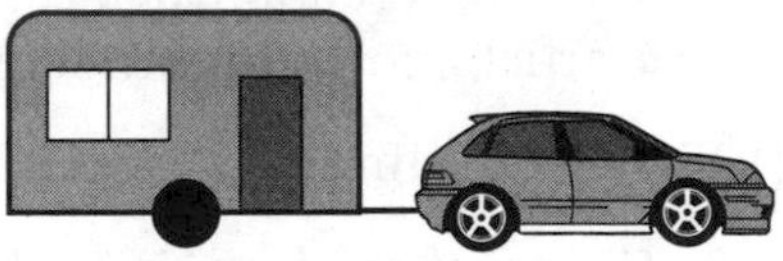

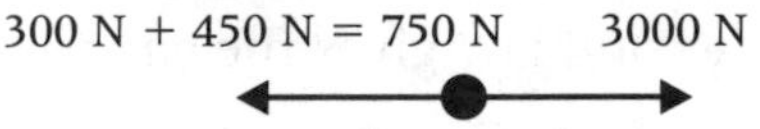

Car and caravan modelled as a particle

Applying Newton's second law gives:

$$2250 = 2000a$$
$$a = 1.125 \text{ m s}^{-2}$$

Now consider the caravan.

The resultant force is $T - 450$, so applying Newton's second law gives the equation

$$T - 450 = 900 \times 1.125$$
$$T = 1463 \text{ N}$$

EXERCISE 6A

1 Two particles, of mass 3 kg and 5 kg, respectively, are attached to the ends of a light, inextensible string that passes over a smooth pulley. Both strings are vertical. Find the acceleration of the particles and the tension in the string.

2 The diagrams below show particles that are connected by a light, inextensible string that passes over a smooth pulley. In each case find the acceleration of the particles and the tension in the string.

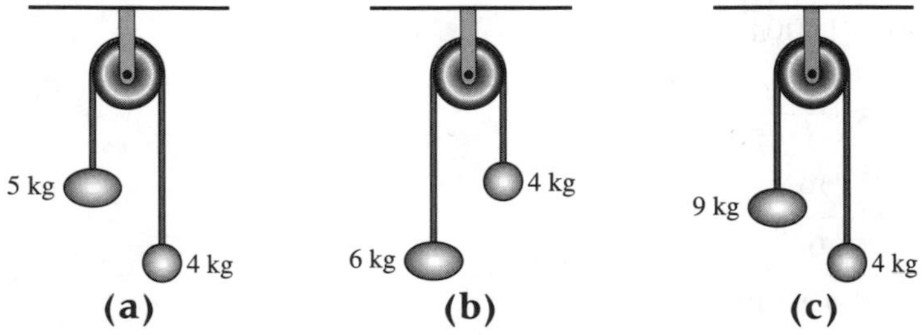

3 Two particles, of mass 4 kg and 7 kg, are attached to the ends of a light inextensible string, which passes over a smooth pulley. Initially the particles are at rest and at the same level. The particles are then released.

(a) Find the acceleration of the particles.

(b) Find the time when the particles are 1 m apart.

4 Two particles of mass m and $\frac{3m}{2}$ are connected by a light, inextensible string that passes over a smooth pulley. Find the acceleration of the particles and the tension in the string in terms of m and g.

5 The diagram shows a block of mass 2 kg, that slides on a rough horizontal table. It is attached by a light, inextensible string to a 3 kg mass. If the block accelerates at 4.9 m s^{-2}, find the coefficient of friction between the block and the table.

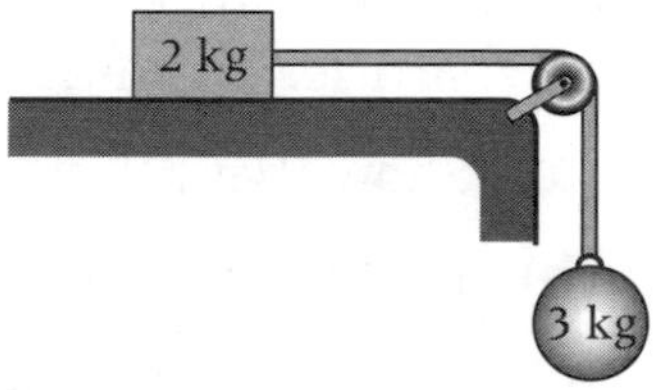

6 A block, of mass 6 kg, rests on a rough, horizontal surface. The coefficient of friction between the block and the surface is 0.2. A light, inextensible, string attached to the block passes over a smooth pulley. A weight, of mass 2 kg, hangs from the other end of the string, as shown in the diagram below.

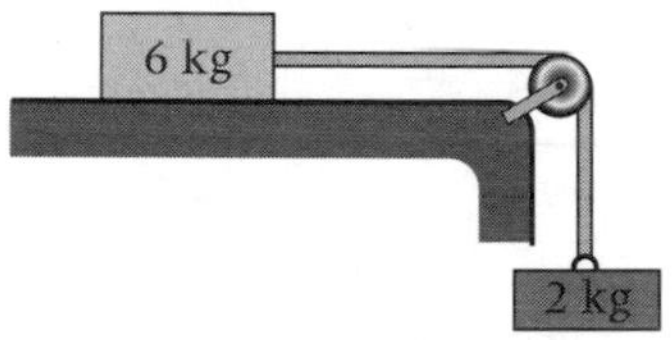

Find the tension in the string and the acceleration of the block. [A]

7 The diagram shows a mass A of 5 kg initially at rest on a horizontal table. A resistance force of 10 N acts against the motion of A which is connected to mass B of 3 kg by a light, inextensible string. The system is released from rest.

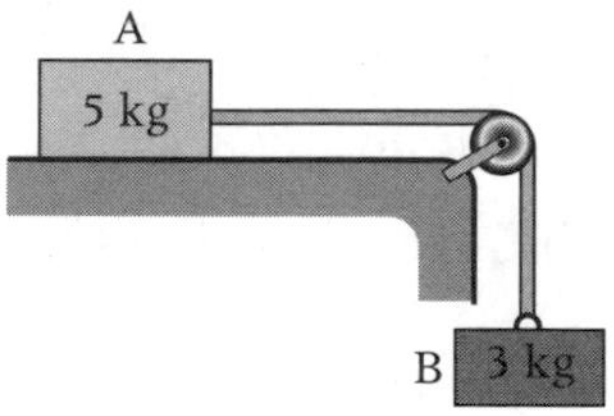

(a) Calculate the acceleration of A.

(b) Calculate the tension in the string.

After a short time, B reaches the floor.

(c) Calculate the acceleration of A now.

8 A breakdown truck tows a car of mass 1200 kg. Assume that the rope is horizontal and that the truck moves on a horizontal surface. Calculate the tension in the tow rope if the car is:

(a) accelerating at 0.5 m s^{-2} and experiencing a resistance force of 500 N

(b) travelling at constant speed but experiencing a resistance force of 400 N.

6

9 Two bodies A and B of mass 4 kg and 2 kg, respectively, are attached by a light inextensible string passing over a smooth pulley. A rests on a table and B hangs over the side. Resistance forces on A amount to 8 N.

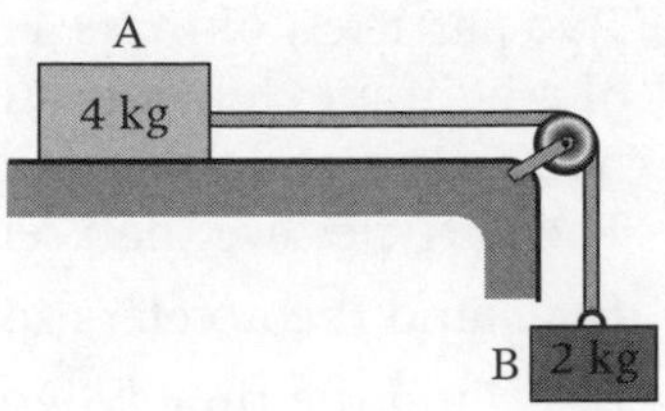

The system is released from rest. Calculate the acceleration of the system and the tension in the string. Find the speed of B when it has fallen 2 m.

10 A particle of mass M is on a rough horizontal table. A light, inextensible string is attached to this particle and passes over a smooth pulley. Attached to the other end of the string, so that it hangs vertically, is another particle of mass m. The coefficient of friction between the particle and the table is μ. The particles are initially at rest.

(a) Show that the acceleration of the particles is $\dfrac{g(m - \mu M)}{m + M}$

(b) Find the tension in the string.

(c) Find the time that it takes the particle on the table to move 0.5 m and its speed at this time.

11 The diagram shows two particles connected by a light inelastic rope. One has a mass of 4 kg and is on a smooth slope and the other has a mass of 6 kg and hangs freely.

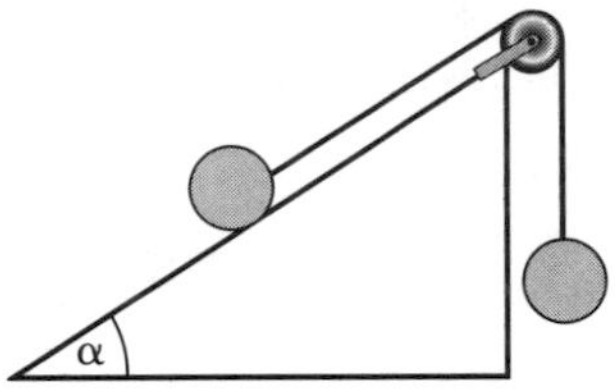

(a) Find the acceleration of the particles if $\alpha = 30°$.

(b) Find the tension in the rope if $\alpha = 10°$.

12 A car of mass 1200 kg is to be pulled 30 m up a slope inclined at 5° to the horizontal. A rope is attached to the car and passes over a smooth pulley and hangs vertically in an old mine shaft. A concrete weight of mass 200 kg is attached to the other end of the rope and released from rest.

(a) Find the acceleration of the car.

(b) How long does the car take to travel the 30 m.

13 At an old mine trolleys move on tracks up a slope inclined at 10° to the horizontal. They are pulled up by a rope that passes over a pulley and then hangs down the mine shaft. A weight of mass 250 kg is attached to this end of the rope. The mass of an empty truck is 800 kg. The truck is subject to a resistance force of 100 N as it moves.

(a) Find the acceleration of an empty truck as it moves up the slope.

(b) At the top of the slope the truck is loaded with 1000 kg of coal. Find the acceleration of the truck as it moves down the slope.

Key point summary

1 Remember that the tension is constant throughout the string. *p86*

2 The magnitude of the acceleration of both particles is the same. *p86*

3 Consider each particle separately, applying Newton's second law to each in turn. *p86*

Test yourself

What to review

If your answer is incorrect – review

1 Particles of mass 4 kg and 3.5 kg are attached to the ends of a light, inextensible string, which passes over a smooth pulley. The system is released from rest. *Section 6.2*

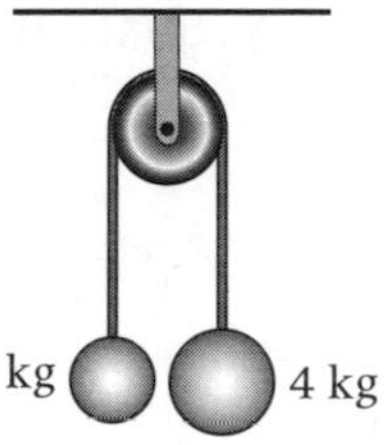

(a) Find the tension in the string.

(b) Find the time that it takes for the heavier particle to fall 0.5 m.

2 A block, of mass 30 kg, rests on a rough horizontal surface. The coefficient of friction between the surface and the block is 0.4. A light inextensible rope is attached to the block and passes over a smooth pulley. A particle, of mass 15 kg, hangs from the other end of the rope. The system is released from rest. *Section 6.2*

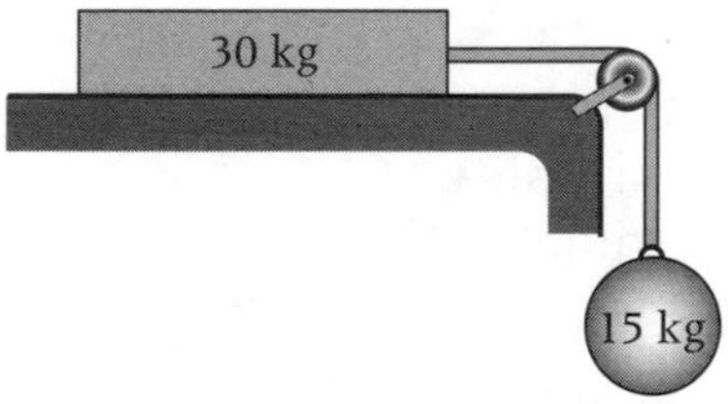

(a) Find the acceleration of the block and the tension in the rope.

(b) Find the speed of the block when it has travelled 40 cm.

6

Test yourself ANSWERS

1 **(a)** 33.6 N. **(b)** 1.24 s.

2 **(a)** 0.653 m s^{-2}, 137.2 N. **(b)** 0.723 m s^{-1}.

CHAPTER 7

Projectiles

Learning objectives

After studying this chapter you should be able to:

- find the time of flight of a projectile
- find the range of a projectile
- find the maximum height of a projectile
- modify equations to take account of height of release
- understand how range varies with the angle of projection
- know that the maximum range is achieved when the angle of projection is 45° provided that the release and landing points are at the same level
- eliminate time from the expressions for x and y to find the equation of the path of a projectile.

7.1 Introduction

In this chapter we will study the motion of objects that move only under the influence of gravity. For example we could consider the paths of a golf ball, a tennis ball, a football or a stunt motorcyclist, while they are in the air.

7.2 A model for the motion of a projectile

We make a number of assumptions that allow us to model these situations. The key assumptions are listed below:

- The object that is moving is modelled as a particle.
- Gravity is the only force that acts, so that we ignore forces such as air resistance and lift that may be present in reality.
- The object is set in motion so that it has a specific initial velocity.

To develop a mathematical model for a projectile we will first consider the motion of a particle that is launched on horizontal ground and is initially at ground level.

As the particle moves only under the influence of gravity it will have a constant acceleration of $g\ \text{m s}^{-2}$. If we define unit vectors $\mathbf{i}$ and $\mathbf{j}$ as horizontal and vertical, respectively, we can write the acceleration as

$$\mathbf{a} = -g\mathbf{j}.$$

We need to be able to describe the initial velocity of the particle. If it initially moves at $V\,\text{m}\,\text{s}^{-1}$, at an angle θ above the horizontal, then the initial velocity of the particle will be

$$\mathbf{u} = V\cos\theta\,\mathbf{i} + V\sin\theta\,\mathbf{j}$$

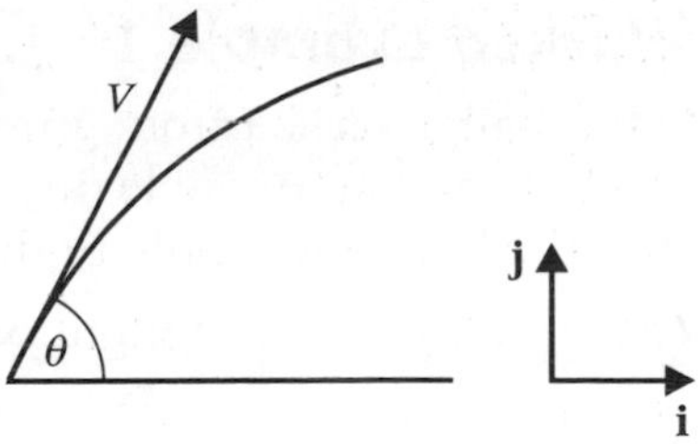

> Using the constant acceleration equations
>
> $$\mathbf{v} = \mathbf{u} + \mathbf{a}t \quad \text{and} \quad \mathbf{r} = \mathbf{u}t + \frac{1}{2}\mathbf{a}t^2$$
>
> allows us to formulate expressions for the velocity and position of the projectile at time t.

$$\begin{aligned}\mathbf{v} &= \mathbf{u} + \mathbf{a}t\\ &= V\cos\theta\mathbf{i} + V\sin\theta\mathbf{j} - gt\mathbf{j}\\ &= V\cos\theta\mathbf{i} + (V\sin\theta - gt)\mathbf{j}\end{aligned}$$

and

$$\begin{aligned}\mathbf{r} &= \mathbf{u}t + \frac{1}{2}\mathbf{a}t^2\\ &= (V\cos\theta\mathbf{j} + V\sin\theta\mathbf{j})t - \frac{1}{2}gt^2\mathbf{j}\\ &= V\cos\theta\, t\mathbf{i} + \left(V\sin\theta\, t - \frac{1}{2}gt^2\right)\mathbf{j}\end{aligned}$$

The diagram below shows the typical path of a projectile.

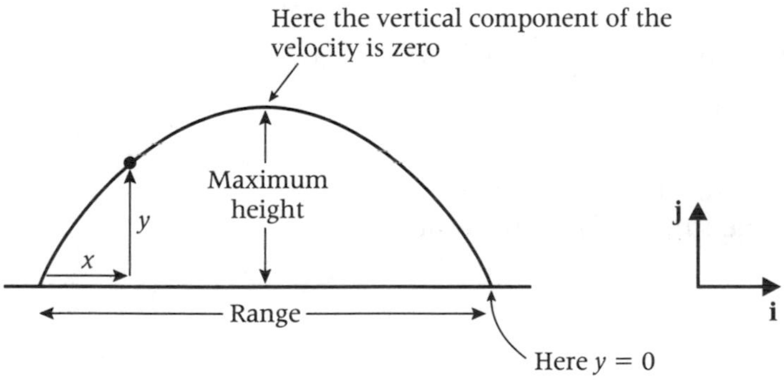

> When working with projectile problems it is often necessary to consider either the horizontal or vertical components of the position. These can be written as
>
> $$x = V\cos\theta\, t \quad \text{and} \quad y = V\sin\theta\, t - \frac{1}{2}gt^2$$

A great many projectile problems can be solved by starting with these two equations. You should learn these formulae so that you can apply them quickly and easily. Their use is illustrated in the following examples.

Worked example 1

A football is kicked from ground level so that its initial speed is $20\ \text{m s}^{-1}$ and at an angle of 30° above the horizontal. Assume that the ball moves only under the influence of gravity.

(a) Find the time of flight of the ball.

(b) Find the range of the ball.

(c) The maximum height of the ball.

Solution

The position of the ball, at time t seconds, is given by:

$$x = V\cos\theta t \qquad y = V\sin\theta t - \frac{1}{2}gt^2$$

$$= 20\cos 30°t \quad \text{and} \quad = 20\sin 30°t - \frac{1}{2}\times 9.8t^2$$

$$= 20\frac{\sqrt{3}}{2}t \qquad = 20\times\frac{1}{2}t - 4.9t^2$$

$$= 10\sqrt{3}t \qquad = 10t - 4.9t^2$$

(a) The time of flight is the time that the ball is in the air, from the time it is kicked to the time when it first hits the ground. This can be found by finding the times when $y = 0$.

$$10t - 4.9t^2 = 0$$

$$t(10 - 4.9t) = 0$$

$$t = 0 \text{ or } t = \frac{10}{4.9} = 2.04 \text{ seconds}$$

The ball is at ground level when $t = 0$ and when $t = 2.04$, so the time of flight is $\frac{10}{4.9} = 2.04$ seconds.

(b) The range is the horizontal distance travelled by the ball. This can be found by substituting the time of flight into the expression for x.

$$x = 10\sqrt{3}\times\frac{10}{4.9} = 35.3\text{ m}$$

(c) The maximum height is attained when the vertical component of the velocity of the ball is zero.

$$0 = V\sin\theta - gt$$

$$0 = 10 - 9.8t$$

$$t = \frac{10}{9.8}$$

Using this value for t gives:

$$y = 10\times\frac{10}{9.8} - 4.9\left(\frac{10}{9.8}\right)^2 = 5.10\text{ m}$$

Including the height of release

A projectile will not always be launched from ground level.

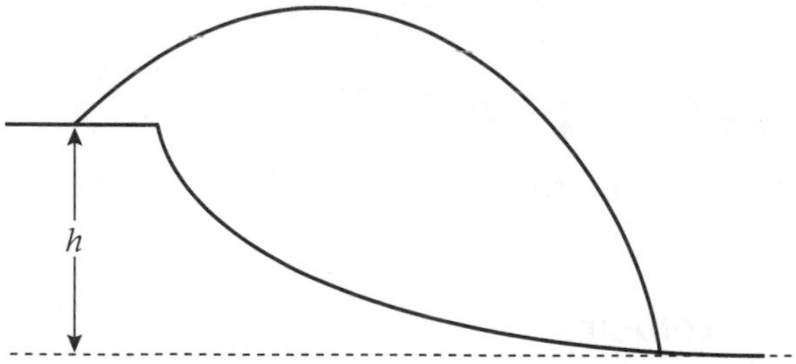

> The expression for y must be modified to take account of this height of release. If the height of release is h m, then
>
> $$Y = Vt \sin\theta - \frac{1}{2}gt^2 + h$$

The next example illustrates a case where the height of release must be included.

Worked example 2

A basket ball is thrown from a height of 1 m at a speed of 10 m s^{-1} and at an angle of 40° above the horizontal. The ball passes through the basket, which is at a height of 3 m above ground level. Find the horizontal distance of the basket from the point where the ball was thrown.

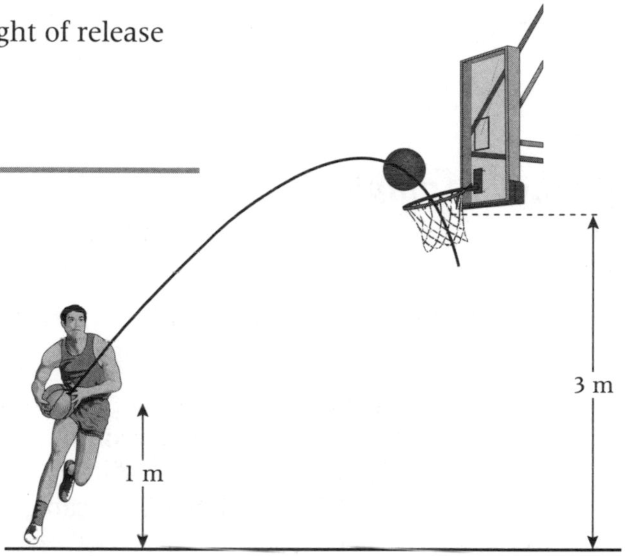

Solution

The position of the ball, at time t seconds, is given by:

$$x = V\cos\theta\, t \qquad \text{and} \qquad y = Vt\sin\theta - \frac{1}{2}gt^2 + h$$

$$= 10\cos 40°t \qquad\qquad = 10\sin 40°t - 4.9t^2 + 1$$

The basket is at a height of 3 m, so when the ball enters the basket $y = 3$. This gives the quadratic equation:

$$3 = 10\sin 40°t - 4.9t^2 + 1$$

$$4.9t^2 - 10\sin 40°t + 2 = 0$$

Solving this quadratic equation gives the two times when the ball will be at a height of 3 m.

$$t = \frac{10\sin 40° \pm \sqrt{100\sin^2 40° - 4 \times 4.9 \times 2}}{2 \times 4.9}$$

$$= 0.5074 \text{ or } 0.8044 \text{ seconds (to four significant figures)}$$

As the ball moves up and then down, it will pass downwards through the basket at the later time. Taking this time and substituting into the expression for x gives:

$$x = 10\cos 40° \times 0.8044$$
$$= 6.16\text{ m (to three significant figures)}$$

So the basket is at a horizontal distance of 6.16 m from the point where the ball was thrown.

Working with the components of the velocity

In some problems the initial velocity of the projectile may not be known, or may be specified in terms of its horizontal and vertical components. In these cases the initial velocity can be expressed as

$$\mathbf{u} = U\mathbf{i} + V\mathbf{j}$$

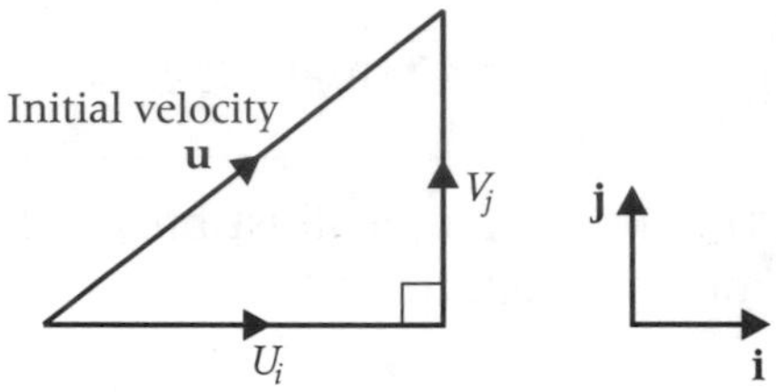

so that the expressions for x and y become

$$x = Ut \quad \text{and} \quad y = Vt - \frac{1}{2}gt^2$$

Worked example 3

An arrow is fired from a bow. It hits a target at the same level at a distance of 91 m from where it was fired and after 1.4 seconds.

(a) Find the horizontal and vertical components of the initial velocity.

(b) A cable passes over the area where the arrow was fired. It is at a height of 6 m and is at a horizontal distance of 20 m from where the arrow was fired. The arrow is fired at a height of 1 m, find the height of the cable above the arrow, when it passes beneath the cable.

Solution

The position of the arrow is given by:

$$x = Ut \quad \text{and} \quad y = Vt - 4.9t^2$$

where U and V are the horizontal and vertical components of the initial velocity, respectively.

(a) First use $x = 91$ and $t = 1.4$, to find U.

$$91 = U \times 1.4$$
$$U = 65\text{ m s}^{-1}$$

Then use $y = 0$ and $t = 1.4$, to find V.

$$0 = V \times 1.4 - 4.9 \times 1.4^2$$
$$V = 4.9 \times 1.4 = 6.86\text{ m s}^{-1}$$

(b) First revise the expression for y to take account of the height at which the arrow was fired. The revised expression is

$$y = Vt - 4.9t^2 + 1$$

The arrow passes under the cable when $x = 20$. This can be used to find the time when the arrow is directly below the cable.

$$20 = Ut$$

$$20 = 65t$$

$$t = \frac{20}{65} = \frac{4}{13} \text{ seconds}$$

This value of t can then be substituted into the revised expression for y to give the height of the arrow when it is below the cable.

$$y = 6.86 \times \frac{4}{13} - 4.9 \times \left(\frac{4}{13}\right)^2 + 1 = 2.65 \text{ m}$$

(to three significant figures)

So the distance between the arrow and the cable is given by

$$6 - 2.65 = 3.35 \text{ m}$$

The arrow passes 3.35 m below the cable.

EXERCISE 7A

1 A rugby ball is kicked from ground level so that its initial velocity is 18 m s^{-1} and at an angle of 45° above the horizontal. Find:

(a) the time of flight

(b) the range

(c) the maximum height of the ball.

2 David kicks a ball, from ground level, with a speed of 20 m s^{-1} at an angle of 30° to the horizontal. How far away from him does the ball land?

3 A bushbaby makes hops with a take-off speed of 6 m s^{-1} and at an angle of 30° to the horizontal. How far does it go in each hop?

4 A ball is thrown with initial speed 20 m s^{-1} at an angle of 60° to the horizontal. Assume that the ball is initially at ground level. How high does it rise? How far has it then travelled horizontally?

5 An athlete launches a shot from a height of 2 m with an initial speed of 10 m s^{-1} and at an angle of 40° above the horizontal. Find the range of the shot. If the throw took place in an indoor arena, find the minimum height of the roof.

6 A locust can make long jumps of 0.7 m at a takeoff angle of 55°. Find its takeoff speed and the maximum height it reaches. (The takeoff speed of locusts is observed to be about 3.4 m s^{-1}, which is higher than the value found using the projectile model. Why do you think this should be?)

7 The horizontal and vertical components of the initial velocity of a cricket ball are 20 m s^{-1} and 25 m s^{-1}. Assume that $g = 10$ m s^{-2} and find the range of the projectile on a horizontal surface. If the ball is caught when it is at a height of 1.5 m and travelling downwards, find the time of flight and the horizontal distance travelled by the ball.

8 A stone is thrown up at an angle of 30° to the horizontal with a speed of 20 m s^{-1} from the edge of a cliff 15 m above sea level so that the stone lands in the sea. Find how long the stone is in the air and how far from the base of the cliff it lands.

9 A stunt motorcyclist takes off at a speed of 35 m s^{-1} up a ramp of 30° to the horizontal to clear a river 50 m wide. Does the cyclist succeed in doing this? Does the motorcyclist have to worry about air resistance?

10 Karen is standing 4 m away from a wall which is 2.5 m high. She throws a ball at 10 m s^{-1} at an angle of 40° to the horizontal from a height of 1 m above the ground. Will the ball pass over the wall?

11 A stone is thrown with speed 10 m s^{-1} at an angle of projection of 30° from the top of a cliff and hits the sea 2.5 s later. How high is the cliff? How far from the base of the cliff does the stone hit the water?

12 A ball is thrown with a speed of 12 m s^{-1} at an angle of 30° to the horizontal. It is initially at ground level. Find the maximum height to which it rises, the time of flight and the range. Find also the speed and direction of flight of the ball after 0.5 s and 1.0 s.

13 A bowler releases a cricket ball from a height of 2.25 m above the ground so that initially its path is level. Find the speed of delivery if it is to hit the ground a horizontal distance of 16 m from the point of release.

14 A tennis ball is hit so that it initially moves horizontally at V m s^{-1} from a point at a height of 2 m. The net which has a height of 0.9 m is at a horizontal distance of 12 m from the point where the ball was hit. Find the minimum value of V for which the ball clears the net. If $V = 30$ m s^{-1} describe what happens to the ball when it has travelled 12 m horizontally.

15 In the Pony Club gymkhana Carol wants to release a ball to drop into a box. The height above the box from which she drops the ball is 1.5 m and the pony's speed is $12\ \text{m s}^{-1}$. How far from the box should Carol drop the ball?

16 A tennis player plays a ball with speed $20\ \text{m s}^{-1}$ horizontally straight down the court from the backline. What is the least height at which she can play the ball to clear the net? How far behind the net does the ball land when it is played at this height?

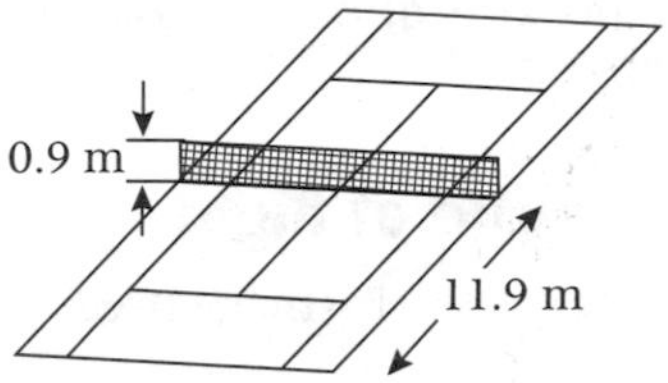

17 A golfer hits a golf ball so that it moves with an initial speed of $40\ \text{m s}^{-1}$ at an angle of 20° above the horizontal.

(a) State two essential assumptions that you should make if you are to estimate the horizontal distance between the point where the ball was hit and the point where it hits the ground for the first time.

(b) Find this distance to the nearest metre.

(c) The ball is actually hit on a raised area that is higher than the ground where the ball lands for the first time. How would this affect the answer that you obtained in part **(b)**? [A]

18 Take $g = 10\ \text{m s}^{-2}$ in this question.

When a tennis ball is served, it is hit from the base-line *A*. It must pass over the net, at *B*, and land **between** the net and service line on the other side, at *C*. The diagram shows the positions of these lines and the height of the net.

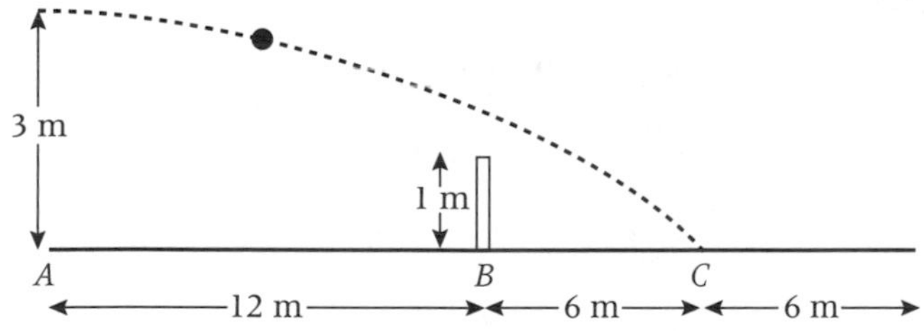

A tennis ball is served so that it initially moves horizontally at a speed of $v\ \text{m s}^{-1}$ from a point 3 m above the base-line *A*. Assume that there are no resistance forces and that the ball moves in a vertical plane at right angles to the net.

(a) (i) Show that v must be greater than $\sqrt{360}\ \text{m s}^{-1}$ if the ball is to go over the net.

(ii) What is the maximum value of v if the ball is to land in the area between *B* and *C*?

(b) A tennis player hits a tennis ball so that is initially moves at a speed of $30\ \text{m s}^{-1}$ and at angle of 5° below the horizontal. Is this serve successful? [A]

7.3 General results for projectiles

It is possible to derive general results for the time of flight, the range and the maximum height of a projectile, launched at ground level on a horizontal surface. These can be used to make predictions about how to obtain the maximum range of a projectile.

The time of flight

We begin with the expression

$$y = V\sin\theta t - \frac{1}{2}gt^2$$

To find the time of flight we must substitute $y = 0$ and solve for t.

$$0 = V\sin\theta t - \frac{1}{2}gt^2$$

$$0 = t(V\sin\theta - \frac{1}{2}gt)$$

$$t = 0 \text{ or } V\sin\theta - \frac{1}{2}gt = 0$$

Solving the second of these equations gives the time of flight.

$$V\sin\theta - \frac{1}{2}gt = 0$$

$$t = \frac{2V\sin\theta}{g}$$

So the time of flight is $\dfrac{2V\sin\theta}{g}$

The range

The range can be found by substituting the time of flight into the expression for x.

$$\begin{aligned} x &= V\sin\theta t \\ &= V\sin\theta \times \frac{2V\sin\theta}{g} \\ &= \frac{V^2 2\cos\theta\sin\theta}{g} \end{aligned}$$

This can be simplified using the double angle formula $\sin 2\theta = 2\cos\theta\sin\theta$, to give

$$x = \frac{V^2\sin 2\theta}{g}$$

So the range of the projectile is $\dfrac{V^2\sin 2\theta}{g}$. From this expression we can consider how the range of the projectile varies with the angle of projection, θ.

The graph shows how $\sin 2\theta$ varies with θ. Note that it takes its maximum value when $\theta = 45°$. This angle of projection will give the maximum range of the projectile.

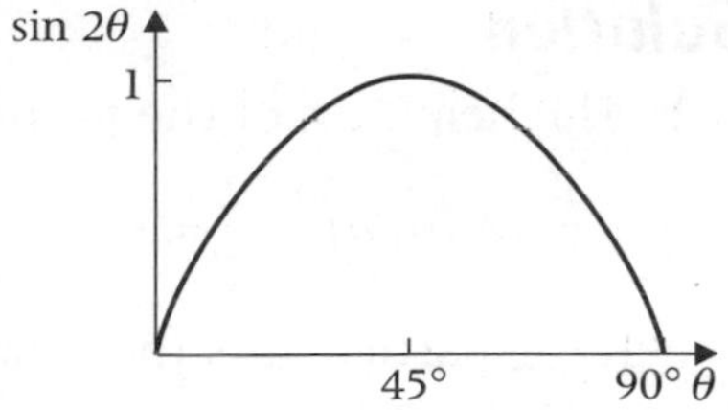

Also note the symmetry of the graph. The value of $\sin 2\theta$ will be the same for 40° and 50°. This means that a particle projected at an angle of 40° will have the same range as a particle projected at 50°. They will not however follow the same path. In general a particle projected at an angle θ will have the same range as a particle projected with the same speed at angle $90 - \theta$. The particle projected at the larger angle will have a higher path and a longer time of flight.

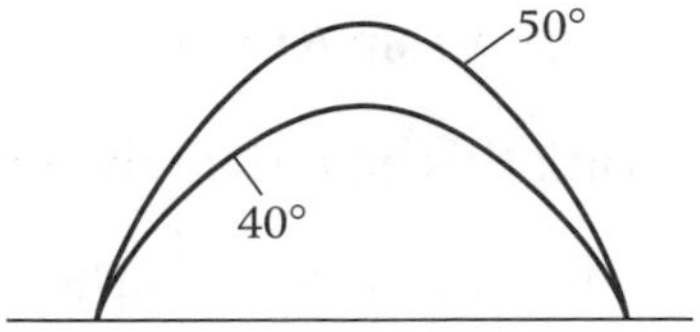

The maximum height

The particle will reach its maximum height when the vertical component of its velocity is zero. That is when

$$V \sin \theta - gt = 0$$

$$t = \frac{V \sin \theta}{g}$$

Substituting this value of t into the expression for y will give the maximum height.

$$y = V \sin \theta t - \frac{1}{2} g t^2$$

$$= V \sin \theta \times \frac{V \sin \theta}{g} - \frac{1}{2} g \left(\frac{V \sin \theta}{g} \right)^2$$

$$= \frac{V^2 \sin^2 \theta}{2g}$$

It is important that you are able to derive these formulae, but you should not simply quote these formulae in an examination.

Worked example 4

A particle is projected with an initial speed of 14 m s^{-1}, from ground level on a horizontal surface.

(a) What are the two possible angles of projection, if the range of the projectile is 10 m?

(b) What is the maximum possible range of a particle projected at this speed?

Solution

(a) The height, y, of the particle at time t is given by

$$y = 14 \sin \theta t - \frac{1}{2}gt^2$$

When the particle hits the ground, its height is zero so we obtain the equation below.

$$0 = 14 \sin \theta t - \frac{1}{2}gt^2$$

Solving this equation gives two values of t.

$$0 = 14 \sin \theta t - \frac{1}{2}gt^2$$

$$0 = t\left(14 \sin \theta - \frac{1}{2}gt\right)$$

$$t = 0 \quad \text{or} \quad t = \frac{28 \sin \theta}{g}$$

The second of these values gives the time of flight. The horizontal distance travelled by the particle is given by

$$x = V \cos \theta t.$$

Substituting $x = 10$, $V = 14$ and the time of flight obtained above gives the equation below.

$$10 = 14 \cos \theta \times \frac{28 \sin \theta}{9.8}$$

$$10 = 40 \cos \theta \sin \theta$$

$$\frac{1}{2} = \sin 2\theta$$

Solving this trigonometric equation then gives the two possible values of θ as 15° and 75°.

(b) The maximum range will be obtained when $\theta = 45°$. This can be substituted into the expression for the time of flight obtained above.

$$t = \frac{28 \sin 45°}{g}$$

$$= \frac{28}{g\sqrt{2}}$$

This can then be substituted into the expression for x to find the range.

$$x = 14 \cos 45° \times \frac{28}{g\sqrt{2}}$$

$$= \frac{14}{\sqrt{2}} \times \frac{28}{g\sqrt{2}}$$

$$= \frac{196}{g}$$

$$= 20 \text{ m}$$

Worked example 5

A particle is projected, so that the horizontal and vertical components of its initial velocity are U and $2U$, respectively. Find the range of this particle in terms of U and g.

Solution

The height, y, of the particle will be given by:

$$y = 2Ut - \frac{1}{2}gt^2.$$

The time of flight can be found by considering when y is zero, which gives the equation below.

$$0 = 2Ut - \frac{1}{2}gt^2$$

$$0 = t\left(2U - \frac{1}{2}gt\right)$$

$$t = 0 \quad \text{or} \quad t = \frac{4U}{g}$$

The horizontal distance travelled by the particle, x, is given by:

$$x = Ut$$

The time of flight can be substituted into this expression to give the range.

$$x = U \times \frac{4U}{g}$$

$$= \frac{4U^2}{g}$$

EXERCISE 7B

1 A projectile, given an initial speed of 20 m s^{-1}, travels a horizontal distance 30 m. What are its possible angles of projection?

2 Robin Hood shoots an arrow with a speed of 60 m s^{-1} to hit a mark on a tree 60 m from him and at the same level as the arrow is released from. What are his possible angles of projection and which one is he likely to choose?

3 Assume $g = 10\text{ m s}^{-2}$ in this question.

A ball is thrown from a point O with a speed 10 m s^{-1} at an angle θ to the horizontal. Show that, if it returns to the ground again at a distance from O greater than 5 m, then θ lies between 15° and 75° whilst the time of flight is between 0.52 s and 1.93 s.

4 The initial speed of a projectile is 40 m s^{-1}. Find the two angles of projection which give a range of 30 m and the times of flight for each of these angles. What is the maximum range that can be achieved?

5 What is the least speed of projection with which a projectile can achieve a range of 90 m? What is the time of flight for this speed?

6 A projectile has range 100 m and reaches a maximum height of 20 m. What is its initial speed and angle of projection?

7 A particle projected from a point O on level ground first strikes the ground again at a distance $4a$ from O after time T. Find the horizontal and vertical components of its initial velocity. [A]

8 A ball is thrown so that it goes as high as it goes forward. At what angle is it thrown?

9 A particle P is projected at time $t = 0$ in a vertical plane from a point O with speed u at an angle α above the horizontal. Write down expressions for the horizontal and vertical components of

(a) the velocity of P at time t

(b) the displacement, at time t, of P from O.

Given that the particle strikes the horizontal plane through O at time T show that

$$T = \frac{2u \sin \alpha}{g}$$

Find, in terms of g and T, the maximum height that P rises above the horizontal plane through O. [A]

10 A particle is projected from ground level, on a horizontal surface. Assume that gravity is the only force acting on it after it has been projected. Its initial velocity has horizontal component U and vertical component V.

(a) Find the range of the particle and show that its maximum height is $\frac{V^2}{2g}$.

(b) Describe how the range would change if:

(i) the horizontal component of the initial velocity was doubled

(ii) both the horizontal and vertical components of the initial velocity were doubled.

(c) If $V = 20 \text{ m s}^{-1}$, find the time for which the height of the particle is greater than half of its maximum height. [A]

11 A projectile is launched at ground level with speed V at an angle θ above the horizontal.

(a) Show that the range of the projectile is $\frac{V^2 \sin(2\theta)}{g}$, and state any essential assumptions that you have to make to obtain this result.

(b) It is estimated that a good shot-putter tries to launch the shot at an angle of 45° above the horizontal, at a speed of 13 m s^{-1} and from a height of 2 m. Find the range of a shot launched in this way.

(c) Also find the range of the shot of part **(b)**, but ignoring the height of release and using the formula found in part **(a)**. State whether or not the formula gives a good prediction, giving a reason for your answer. [A]

12 A golf ball is hit so that its initial velocity is 30 m s^{-1} at 40° above the horizontal.

(a) Show that the ball travels approximately 90 m before it hits the ground, stating any assumptions that you make to obtain this value.

(b) The ball in fact travels a horizontal distance of 100 m. Find the difference in height between the point from where it was hit and the point where it hit the ground.

(c) Show that it would be impossible for a golf ball with an initial speed of 30 m s^{-1} to travel 100 m if there was no difference in the level of the point from where it was hit and the point where it lands. [A]

13 At time $t = 0$ a particle is projected from a point O with speed 49 m s^{-1} and in a direction which makes an acute angle θ with the horizontal plane through O. Find, in terms of θ, an expression for R, the horizontal range of the particle from O.

The particle also reaches a height of 9.8 m above the horizontal plane through O at times t_1 seconds and t_2 seconds. Find, in terms of θ, expressions for t_1 and t_2.

Given that $t_2 - t_1 = \sqrt{17}$ seconds, find θ.

Hence show that $R = \frac{245\sqrt{3}}{2}$. [A]

7.4 The equation of the path of a projectile

It is possible to find the equation of the path of a projectile, by eliminating t from the two equations below.

$$x = V\cos\theta t \quad \text{and} \quad y = V\sin\theta t - \frac{1}{2}gt^2$$

From the first of these equations

$$t = \frac{x}{V\cos\theta}$$

This can then be substituted into the second equation to give

$$y = V\sin\theta \times \frac{x}{V\cos\theta} - \frac{1}{2}g\left(\frac{x}{V\cos\theta}\right)^2$$

$$= x\tan\theta - \frac{gx^2}{2V^2\cos^2\theta}$$

$$= x\tan\theta - \frac{gx^2\sec^2\theta}{2V^2}$$

$$= x\tan\theta - \frac{gx^2(1+\tan^2\theta)}{2V^2}$$

Note: $\sec\theta = \frac{1}{\cos\theta}$ and $\sec^2\theta = 1 + \tan^2\theta$

This approach can be very useful for finding the possible angles of projection, if information is available about a point through which a projectile passes.

Worked example 6

A golf ball is hit so that its initial velocity is 42 m s $^{-1}$. It travels a horizontal distance of 150 m and lands on ground that is 4 m lower than its initial position. Find the possible angles of projection.

Solution

First consider the horizontal motion of the ball. The values $x = 150$ and $V = 42$ can be substituted into $x = V\cos\theta t$. Solving for t then gives the time of flight as below.

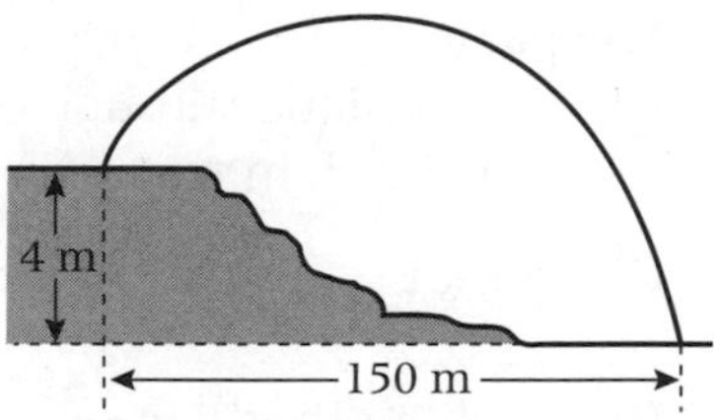

$$150 = 42\cos\theta t$$

$$t = \frac{25}{7\cos\theta}$$

Now consider the vertical motion. The values $V = 42$ and $y = -4$ can be substituted, along with the time of flight into the equation $y = V \sin \theta t - \frac{1}{2}gt^2$ to give the equation below.

$$-4 = 42 \sin \theta \left(\frac{25}{7 \cos \theta}\right) - \frac{1}{2}g\left(\frac{25}{7 \cos \theta}\right)^2$$

$$-4 = 150 \tan \theta - 62.5(1 + \tan^2 \theta)$$

$$0 = 62.5 \tan^2 \theta - 150 \tan \theta + 58.5$$

Solving this quadratic equation gives two values for θ.

$$\tan \theta = \frac{150 \pm \sqrt{150^2 - 4 \times 62.5 \times 58.5}}{2 \times 62.5}$$

$$= 1.9099 \quad \text{or} \quad 0.4901$$

From these values for tan θ we obtain 62.4° or 26.1°.

Worked example 7

A tennis player makes a return at a speed of 14 m s^{-1} and at a height of 3 m. The ball lands in the court at a horizontal distance of 12 m from the player. Find the two possible angles of projection of the ball and state which one the player is more likely to have used.

Solution

Substituting $V = 14$ and $x = 12$ in the expression $x = V \cos \theta t$ gives the time of flight as below.

$$12 = 14 \cos \theta t$$

$$t = \frac{6}{7 \cos \theta}$$

The height of the ball at time t is given by:

$$y = V \sin \theta t - \frac{1}{2}gt^2 + 3$$

Substituting the time of flight for t and the values $y = 0$ and $V = 14$ gives the equation below.

$$0 = 14 \sin \theta \times \frac{6}{7 \cos \theta} - \frac{1}{2}g\left(\frac{6}{7 \cos \theta}\right)^2 + 3$$

$$0 = 12 \tan \theta - 3.6(1 + \tan^2 \theta) + 3$$

$$0 = 3.6 \tan^2 \theta - 12 \tan \theta + 0.6$$

This quadratic equation can now be solved to find two possible values for tan θ.

$$\tan\theta = \frac{12 \pm \sqrt{12^2 - 4 \times 3.6 \times 0.6}}{2 \times 3.6}$$

$$= 3.2826 \quad \text{or} \quad 0.0508$$

These give the possible values of θ as 73.1° or 2.9°. The lower angle would be the more likely as it will make it more difficult for the other player to return the ball.

EXERCISE 7C

1 A ball is kicked with speed 25 m s^{-1} at an angle of projection of 45°. How high above the ground is it when it has travelled 10 m horizontally?

2 One of the Egyptian pyramids is 130 m high and the length of each side of its square base is 250 m. Is it possible to throw a stone with initial speed 25 m s^{-1} from the top of the pyramid so that it strikes the ground beyond the base?

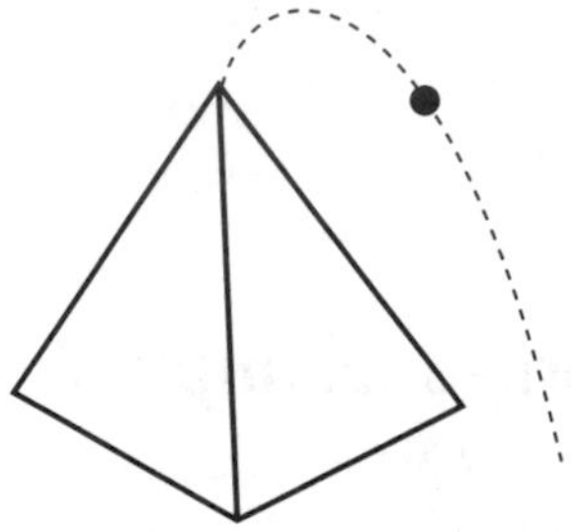

3 Rashid is standing 4 m away from a wall which is 5 m high. He throws a ball at 10 m s^{-1} at an elevation of 40° above the horizontal and at a height of 1 m above the ground. Will the ball pass over the wall?

If he throws at an angle θ, find the inequality that θ must satisfy for the ball to clear the wall.

4 In this question take $g = 10$ m s^{-2}.

A stone is thrown with speed 15 m s^{-1} from a cliff 40 m high to land in the sea at a distance of 30 m from the cliff. Show that there are two possible angles of projection and that they make a right angle.

5 A girl stands on level ground and throws a ball from a height of 1.5 m above the ground so that it just clears two walls. The tops of the walls are 5 m and 7.5 m above the ground and their respective horizontal distances from the point of projection are 8 m and 16 m. Find the speed of projection and the angle of projection of the ball. [A]

6 A golfer hits a ball from a point O with initial speed 28 m s^{-1} in a direction inclined at θ above the horizontal. Given that the ball just clears a fence 17.5 m high, the base of which is a horizontal distance of 60 m from O, show that $3\tan\theta = 4$.

Find the time taken for the ball to reach the fence.

Show that the ball is at least 17.5 m above O for a period of $\frac{18}{7}$ seconds. [A]

7 In this question assume $g = 10\,\text{m s}^{-2}$.

A rugby ball kicked from ground level is to be modelled as a particle that is projected with speed v at an angle α above the horizontal.

(a) Show that the vertical displacement of the ball is given by,

$$y = x \tan \alpha - \frac{gx^2(1 + \tan^2 \alpha)}{2v^2}$$

where x is the horizontal displacement.

(b) State **two** factors that could affect the motion of the ball that are not taken into account in this model.

(c) A rugby player kicks a ball with an initial speed of $20\,\text{m s}^{-1}$. The player hopes that the ball will pass over a bar of height 4 m that is 20 m horizontally from where the ball was kicked. By modelling the ball as a particle find the range of values for the angle of projection for which the ball will go over the bar.

(d) The ball is kicked with the largest possible of these angles of projection. What will happen to the ball?

(e) The ball is in fact kicked while a wind of $5\,\text{m s}^{-1}$ is blowing. A refined model assumes that the horizontal component of the velocity is increased by $5\,\text{m s}^{-1}$. Show that for an angle of 27° the ball will go over the bar. [A]

8 A golf ball is struck from a point O, leaving O with speed $21\sqrt{2}\,\text{m s}^{-1}$. The ball lands, without previously bouncing, at a point A, which is 10 m below the horizontal plane through O. Find the speed of the ball at the instant it lands at A.

The ball just clears the top of a tree which is at a horizontal distance of 72 m from O, the top of the tree being 9 m above the horizontal plane through O.

Given that the ball is projected from O at an elevation θ, show that $\tan \theta$ is equal to either $\frac{3}{4}$ or $\frac{7}{4}$.

Using the same diagram, sketch the path of the ball for each of the possible values of θ. [A]

9 The motion of the ball in a successful free shot in basketball is illustrated opposite.

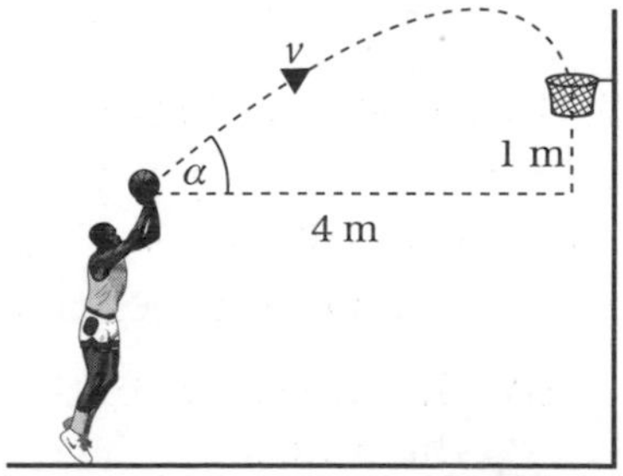

The model assumes that the ball is a point particle, acted on by constant gravity, g.

The ball is projected from a position, distance 4 m horizontally and 1 m vertically from the basket, with speed $v\,\text{m s}^{-1}$ at an angle $\alpha°$ to the horizontal. The ball falls into the basket.

(a) Show that v and α must satisfy

$$1 = 4\tan\alpha - \frac{78.4}{v^2}\sec^2\alpha.$$

(b) Use this equation to find the required speed of projection when angle α equals 45°.

(c) Also, use this equation to find the two possible trajectories when $v = 8.0\ \text{m s}^{-1}$.

For the ball to fall through the basket, the angle made with the vertical at the basket should be as small as possible. Which of your two solutions above would be preferred? [A]

10 A shot putter can release the shot at a height H above the ground with speed U.

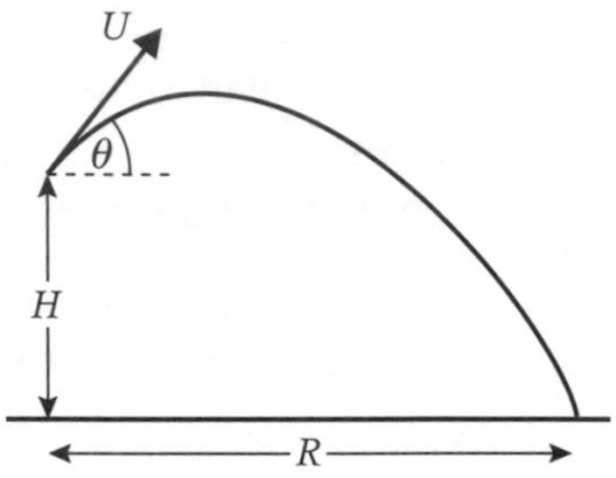

Show that, when the shot is projected at an angle θ to the horizontal, it hits the ground at a horizontal distance R from the point of projection, where R is given by

$$R\tan\theta - g\frac{R^2}{2U^2}\sec^2\theta + H = 0$$

Show that R has a maximum as a function of θ when

$$\tan\theta = \frac{U^2}{gR}.$$

Hence show the maximum range is

$$\frac{U}{g}(U^2 + 2gh)^{\frac{1}{2}}.$$

11 A particle is projected at time $t = 0$ with speed $49\ \text{m s}^{-1}$ at an angle α above the horizontal. The horizontal and vertical displacements from 0, the point of projection, at time t are x m and y m, respectively. Obtain x and y in terms of α, g and t and hence deduce that, when $x = 140$ and $g = 9.8\ \text{m s}^{-2}$, $y = 140\tan\alpha - 40(1 + \tan^2\alpha)$.

Find the numerical values of the constants a and b so that this equation can be re-written as

$$y = a - 40(\tan\alpha - b)^2.$$

The particle has to pass over a wall 20 m high at $x = 140$, find

(a) the value of $\tan\alpha$ such that the particle has the greatest clearance above the wall,

(b) the two values of $\tan\alpha$ for which the particle just clears the wall. [A]

12 A free kick is to be taken at a football match. The ball is placed on the pitch at a distance of 20 m from the goal line. A wall of defenders stand 5 m from the ball and have height 1.8 m. The height of the crossbar of the goal is 2 m. The ball is kicked with an initial velocity of 20 m s^{-1} at an angle α above the horizontal. Find the minimum and maximum values of α for which a goal can be scored, assuming that the ball is a particle and that there is no air resistance.

Key point summary

Formulae to learn

$$x = V \cos \theta t$$

$$y = V \sin \theta t - \frac{1}{2} g t^2$$

- A projectile moves only under the influence of gravity and so it will have an acceleration of $g \text{ m s}^{-2}$. *p94*
- The constant acceleration equations can be used to formulate expressions for velocity and position of a projectile at time t. *p95*
- The horizontal component of velocity of a projectile can be written as $x = V \cos \theta t$ and the vertical component of velocity can be written as $y = V \sin \theta t - \frac{1}{2} g t^2$. *p95*
- The height of release can be calculated by modifying the expression for y. *p97*
- Range will vary with the angle of projection. *p102*
- Maximum range is achieved when the angle of projection is 45°. *p103*
- Time must be eliminated from the expressions of x and y to find the equation of a path of a projectile. *p108*

Test yourself

What to review

If your answer is incorrect – review

1 A particle is projected from ground level on a horizontal plane, with speed 30 m s^{-1} and at an angle of 60° above the horizontal. *Section 7.2*

(a) Find the range and maximum height of the particle.

(b) Revise your answers to **(a)** if the particle was projected at a height of 5 m.

continued overleaf

2 A particle is projected from ground level on a horizontal plane at a speed of 40 m s^{-1}. The range of the particle is 80 m. *Section 7.3*

(a) Find the possible angles of projection of the particle.

(b) Find the maximum possible range of the particle on the horizontal plane.

---

3 A basket ball is thrown from a height of 1.5 m with an initial speed of 12 m s^{-1}. The ball passes through a basket that is at a height of 3 m and a horizontal distance of 4 m from the point where the ball is thrown. *Section 7.4*

(a) Find the possible angles of projection of the basket ball.

(b) Which of these do you think that the player used? Explain your choice.

Test yourself ANSWERS

1 **(a)** 79.5 m, 34.4 m. **(b)** 82.3 m, 39.4 m.

2 **(a)** 14.67° or 75.3°. **(b)** 163 m.

3 **(a)** 28.9° or 81.6°. **(b)** Lower as ball more difficult for other players to intercept.

CHAPTER 8

Momentum

Learning objectives

After studying this chapter you should be able to:

- define the momentum of an object
- understand what happens to momentum of a particle during collision
- use the principle of conservation of momentum to calculate momentum after collision.

8.1 Introduction

In this chapter we will consider a quantity related to the motion of a body called its momentum. The momentum is defined as the product of the mass and the velocity of the body that is moving. In this chapter we will restrict ourselves to motion in one dimension, but the ideas can be extended to more than one dimension. The main focus of the chapter will be to consider what happens to the momentum of bodies that are involved in collisions.

8.2 Momentum

> The momentum of an object is defined as mv where m is the mass and v is the velocity.

Worked example 1

A car has mass 1200 kg and is travelling at 20 m s^{-1}. A lorry has mass 3500 kg and is travelling in the opposite direction to the car at 12 m s^{-1}.

Calculate the momentum of the car and the momentum of the lorry.

Solution

Defining the direction in which the car is travelling as positive gives:

$$\text{Momentum of car} = 1200 \times 20 = 24\,000 \text{ kg m s}^{-1}$$

$$\text{Momentum of lorry} = 3500 \times (-12) = -42\,000 \text{ kg m s}^{-1}$$

Note that the units for momentum are kg m s^{-1} or alternatively N s.

8.3 Conservation of momentum

We will now consider what happens to the momentum of a particle that is involved in a collision.

Consider two particles that are involved in a collision. The diagrams below show the velocities of the particles before and after the collision.

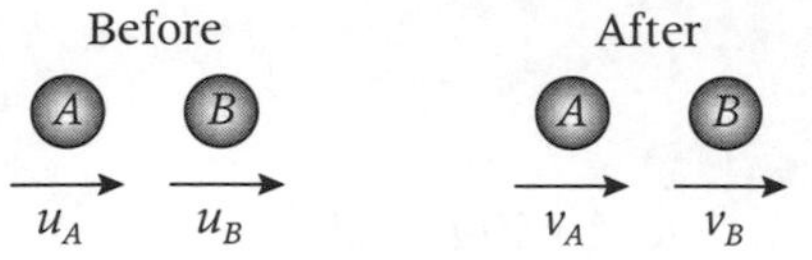

While the particles are in contact they exert equal, but opposite forces on each other. If the magnitude of each of these forces is assumed to be a constant F, then we can find a relationship between the accelerations of the particles during the collision.

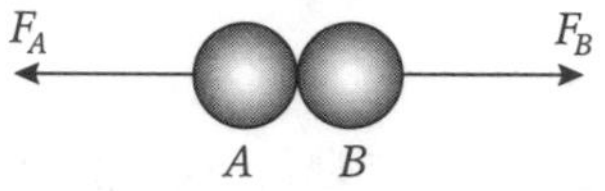

$$-F_A = F_B$$

$$-m_A a_a = m_B a_B$$

As the particles are in contact for the same period of time, t, we can also state that:

$$-m_A a_A t = m_B a_B t$$

The constant acceleration equation $v = u + at$ can be used to eliminate t and the accelerations from our equation. Using $a_A t = v_A - u_A$ and $a_B t = v_B - u_B$. Then substituting into expression (1) gives:

$$m_A a_A t = m_B a_B t$$

$$-m_A(v_A - u_A) = m_B(vb - u_B)$$

$$m_A u_A + m_B u_B = m_A v_A + m_B v_B$$

or

total momentum before collision
= total momentum after collision

> This is called the principle of conservation of momentum. It can be applied to any collision provided that no external forces act during the collision.

Worked example 2

A toy train, of mass 200 g, is moving along a straight track at $1.8\,\text{m s}^{-1}$, when it collides with a stationary truck of mass 300 g, during the collision the trucks are coupled together. Find the speed of the truck after the collision.

Solution

The diagram shows the velocities of the trucks before and after the collision.

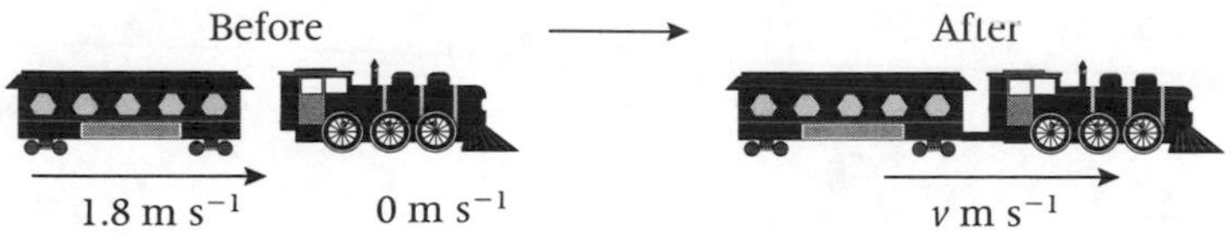

After the collision both trucks have the same speed, v. So applying the principle of conservation of momentum gives:

$$0.2 \times 1.8 + 0.3 \times 0 = (0.2 + 0.3)v$$

$$0.36 = 0.5v$$

$$v = 0.72\ \text{m s}^{-1}$$

Worked example 3

Two particles travel towards each other along a straight line. One has mass 3 kg and speed $4\ \text{m s}^{-1}$. The other has mass 5 kg and speed $2\ \text{m s}^{-1}$. When they collide the 3 kg mass is brought to rest. What happens to the 5 kg mass?

Solution

The diagrams show the velocities of the particles before and after the collision.

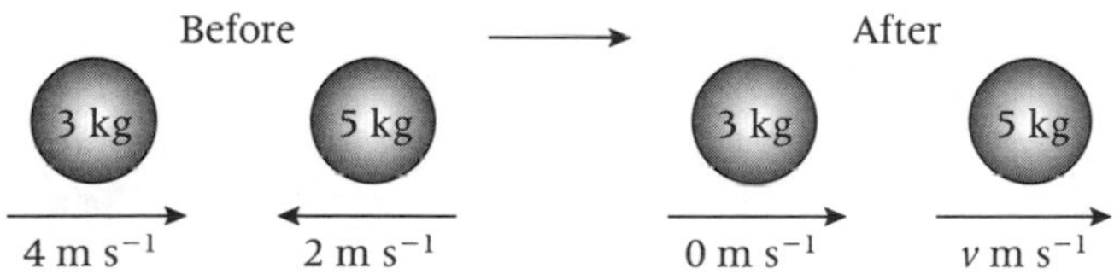

Applying the principle of conservation of momentum gives:

$$3 \times 4 + 5 \times (-2) = 3 \times 0 + 5 \times (-v)$$

$$2 = -5v$$

$$v = -0.4\ \text{m s}^{-1}$$

Worked example 4

A person, of mass 60 kg, stands on a truck of mass 20 kg, that is free to move on a horizontal surface. On the truck there is also a large medicine ball of mass 5 kg. Initially the truck is at rest. The person then throws the ball off the truck so that its initial velocity is $4\ \text{m s}^{-1}$ horizontally. Find the final speed of the truck.

8

Solution

The diagram shows the velocities before and after the throw.

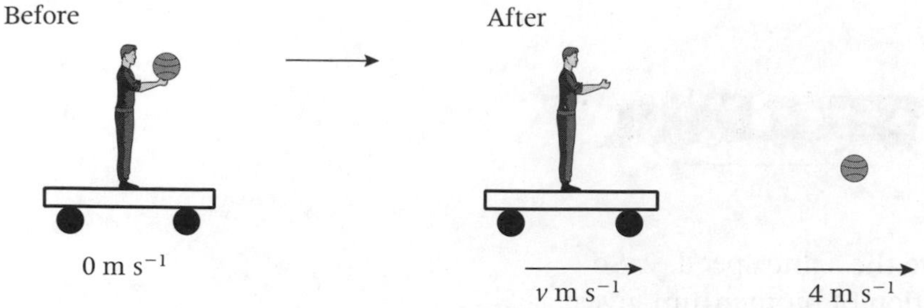

Using the principle of conservation of momentum gives:

$$85 \times 0 = 80v + 5 \times 4$$

$$v = -\frac{20}{80}$$

$$= -0.25 \text{ m s}^{-1}$$

So the truck moves at 0.25 m s^{-1}, after the throw.

EXERCISE 8A

1 Calculate the momentum of:

(a) a train, of mass 120 tonnes, travelling at 40 m s^{-1}

(b) a table tennis ball, of mass 3 g, travelling at 4 m s^{-1}

(c) a car, of mass 1200 kg, travelling at 36 kph.

2 In the diagram A and B have masses 2 kg and 3 kg and move with initial velocities as shown. The collision reduces the velocity of A to 2 m s^{-1}. Find the velocity of B after the collision.

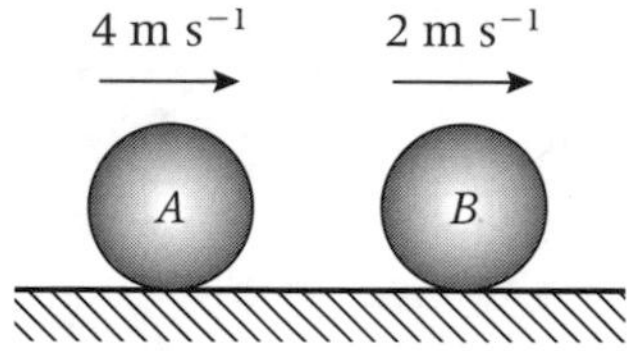

3 The two bodies in the diagram have masses 3 kg and 5 kg, respectively. They are travelling with speeds 4 m s^{-1} and 2 m s^{-1}. The bodies coalesce on impact. Find the speed of this body.

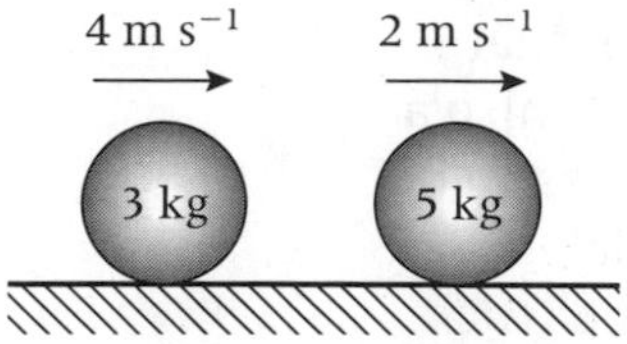

4 A particle A of mass 250 g collides with a particle B of mass 150 g. Initially A has velocity 7 m s^{-1} and B is at rest. After the collision, the velocity of B is 5 m s^{-1}. Calculate the velocity of A after the impact.

5 Two railway trucks, each of mass 8 tonnes, are travelling in the same direction and along the same tracks with velocities 3 m s^{-1} and 1 m s^{-1}, respectively. When the trucks collide they couple together. Calculate the velocity of the coupled trucks.

6 A child of mass 40 kg runs and jumps onto a skateboard of mass 4 kg. If the child was moving forward at 0.68 m s^{-1} when he jumped onto the skateboard, find the speed at which they move.

7 A tow truck of mass 3 tonnes is attached to a car of mass 1.2 tonnes by a rope. The truck is moving at a constant 3 m s^{-1} when the tow rope becomes taut and the car begins to move. Assume that both vehicles move at the same speed once the rope is taut, and find this speed.

8 A red snooker ball that was at rest is hit directly by a white ball moving at 0.8 m s^{-1}. After the collision the red ball moves at 0.75 m s^{-1}. Find the speed of the white ball after the collision.

9 Two identical uniform smooth spheres, A and B, each of mass m, moving in opposite directions with speeds u and $3u$, respectively, collide directly. The sphere B is brought to rest by the collision. Find the speed of A after the collision and state the direction in which it moves.

10 A car of mass 1.2 tonnes collides with a stationary van of mass 2.4 tonnes. After the collision the two vehicles become entangled and skid 15 m before stopping. Police accident investigators estimate that the magnitude of the friction force during the skid was 2880 N. Assume the road is horizontal and that all the motion takes place in a straight line.

(a) Find the speed of the vehicles just after the collision.

(b) Find the speed of the car before the collision. [A]

11 A train of total mass 110 tonnes and velocity 80 kph crashes into a stationary locomotive of mass 70 tonnes.

(a) Calculate the velocity of the combined system immediately after impact.

The trains plough on for a further 40 m.

(b) Calculate the average deceleration and the resistance to motion.

12 A pile-driver consists of a pile of mass 200 kg and a driver of mass 40 kg. The driver drops on the pile with velocity $6\ \text{m s}^{-1}$ and sticks to the top of the pile.

(a) Calculate the velocity of the pile immediately after impact.

Resistances to motion of the pile amount to 1400 N.

(b) Calculate the distance penetrated by the pile.

13 Two uniform smooth spheres, A of mass 0.03 kg and B of mass 0.1 kg, have equal radii and are moving directly towards each other with speeds of $7\ \text{m s}^{-1}$ and $4\ \text{m s}^{-1}$, respectively. The spheres collide directly and B is reduced to rest by the impact. Find the speed of A after the impact. [A]

14 Two cars are initially 36 m apart travelling in the same direction along a straight, horizontal road. The car in front is initially travelling at $10\ \text{m s}^{-1}$, but decelerating at $2\ \text{m s}^{-2}$. The other car travels at a constant $15\ \text{m s}^{-1}$.

(a) Model the cars as particles. By finding the distance travelled by each car after t seconds, show that the distance between the two cars is $36 - 5t - t^2$ metres. Find when they would collide if neither car takes avoiding action.

(b) Would it be necessary to revise your answers to part **(a)** if the cars were not modelled as particles? Give reasons to support your answer.

The mass of the front car is 1500 kg and the mass of the other car is 1000 kg.

(c) The cars do collide and after the collision the two cars move together. Find the speed of the cars just after the collision. [A]

15 In this question take g to be $10\ \text{m s}^{-2}$.

A skier of mass 60 kg is skiing down the slope shown.

(a) The skier is subject to a number of forces. One of these is air resistance. State two factors that would influence the magnitude of this force.

(b) Draw a clear diagram to show **all** the forces on the skier.

(c) Show that the magnitude of the force resisting the skier's motion when he is accelerating at 0.2 m s^{-2} is approximately 143 N, and find the magnitude of the normal reaction on the skier.

(d) In reality the skier's speed does not increase indefinitely, but reaches a maximum of 8 m s^{-1}. Assume that the resistive forces comprise a constant force of magnitude 120 N and an air resistance force that is proportional to the speed of the skier. Find the constant of proportionality.

(e) While moving at a constant speed of 8 m s^{-1} the skier collides with another skier of mass 70 kg moving at a constant 6 m s^{-1} in the same direction. Immediately after the collision they move together in the same direction as they were moving before the collision. Find their speed immediately after the collision. [A]

16 In this question take $g = 10$ m s^{-2}.

A car of mass 1200 kg is travelling up a slope inclined at an angle of 4° to the horizontal. Assume that the car experiences a constant resistive force of magnitude $80v$ N when the car is moving at speed v m s^{-1}.

(a) Model the car as a particle. Draw a diagram to show the forces acting on the car as it travels up the slope with a constant or increasing speed.

(b) At the instant when the car is moving up the slope at 10 m s^{-1} and accelerating at 1.2 m s^{-2}, show that the magnitude of the force acting in the direction of motion is 3077 N to the nearest Newton.

(c) Find the maximum speed of the car up the slope if the magnitude of the forward force remains constant at 3077 N.

(d) At the top of the slope the car moves onto a level surface, travelling at 20 m s^{-1} and collides with a stationary car. After the collision the two cars move together at 12 m s^{-1}. Find the mass of the stationary car. [A]

Key point summary

1 Momentum is the product of mass and velocity, that is momentum $= mv$ p115

2 Momentum is conserved in collisions, when no external forces act. p116

3 Conservation of momentum p116

$$m_A u_A + m_B u_B = m_A v_A + m_B v_B$$

Test yourself	What to review
	If your answer is incorrect – review
1 A car, of mass 1000 kg, is travelling at $20\ \text{m s}^{-1}$, when it drives into a truck, of mass 4000 kg, which was moving at $10\ \text{m s}^{-1}$. After the collision the two vehicles move together. Find the speed of the vehicles after the collision.	*Section 8.3*
2 A child of mass 40 kg stands on a skate board, of mass 3 kg. Initially both are at rest. The boy jumps off so that he travels horizontally at $1.2\ \text{m s}^{-1}$. Find the speed of the skateboard.	*Section 8.3*
3 Two particles are travelling towards each other when they collide. One has mass 2 kg and was travelling at $5\ \text{m s}^{-1}$ before the collision and the other has mass 3 kg and a velocity of $6\ \text{m s}^{-1}$ before the collision. The 2 kg mass reverses direction and moves at $1\ \text{m s}^{-1}$ after the collision. Describe how the 3 kg mass moves after the collision.	*Section 8.3*

Test yourself ANSWERS

1 $12\ \text{m s}^{-1}$.

2 $16\ \text{m s}^{-1}$.

3 $2\ \text{m s}^{-1}$ in same direction as before collision.

CHAPTER 9

Moments and centres of mass

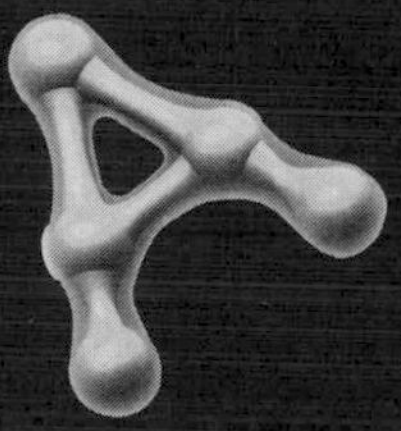

Learning objectives

After studying this chapter you should be able to:

- find the moment of a force
- know that the resultant force and moment must both be zero for a rigid body to be in equilibrium
- find the centre of mass of a system of particles or of a composite body
- know that, when a body is suspended in equilibrium from a point, the centre of mass is directly below the point of suspension.

9.1 Introduction

In this chapter we move on from modelling objects as particles and introduce the idea of a rigid body. This is a body that does not change shape, but that does have size. The major difference between a body modelled as a particle and a rigid body is that the place where the force acts on a rigid body becomes important.

Consider two forces of equal magnitude, but opposite directions that act on a book resting on a table. The diagram shows the particle model and a possible rigid body model.

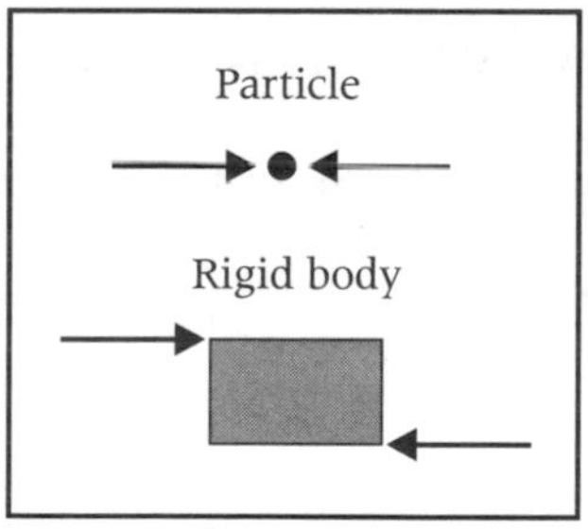

The two situations shown above would produce different results if the forces were applied as shown. The particle would remain at rest, but the rigid body would rotate. It is the way in which the forces are applied that causes the body to rotate. In the first section of this chapter we will introduce the idea of the moment of a force, which gives a measure of the turning effect of a force.

9.2 The moment of a force

The turning effect of a force depends on the size and direction of the force and the point where the force acts. Consider a door that is hinged at O, as shown in the diagram.

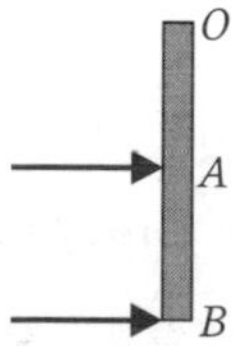

It is easier to open the door with a force applied at B than at A. The greater the distance of the force from the hinge the greater the turning effect.

> The moment of a force about a point is defined as the magnitude of the force multiplied by the perpendicular distance from the point to the force or the line of action of the force.

In this diagram the distance specified is perpendicular to the force and so the perpendicular distance is simply d.

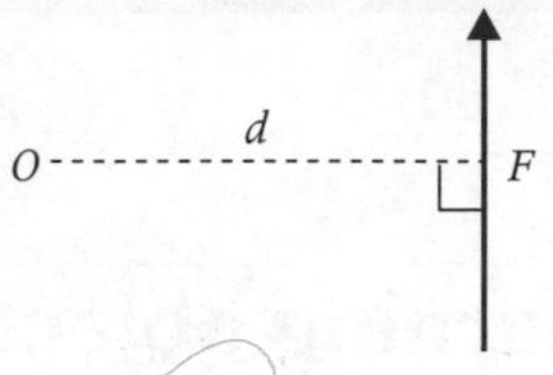

Moment of the force about $O = Fd$

If the distance specified is not perpendicular to the force, then the perpendicular distance must be calculated.

In this case the specified distance, d, is not the perpendicular distance and so this must be calculated. Here the perpendicular distance is $d \sin\theta$, so the moment about O is

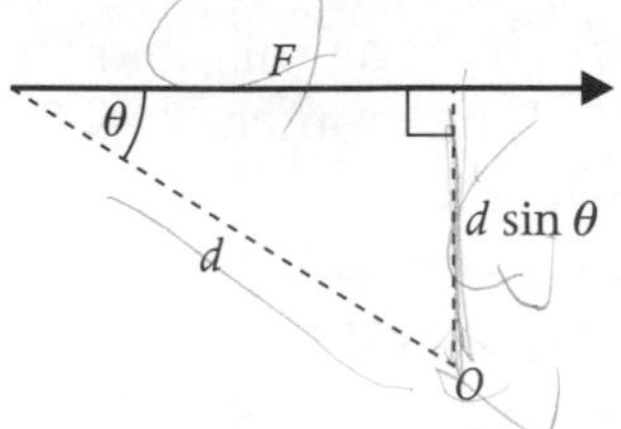

$$Fd \sin\theta$$

Sometimes a force has to be extended, to show the line of action of the force, so that the perpendicular distance can be found.

> A moment can have a clockwise or anticlockwise turning effect. An anti-clockwise moment is defined as a positive moment and a clockwise moment is defined as a negative moment.

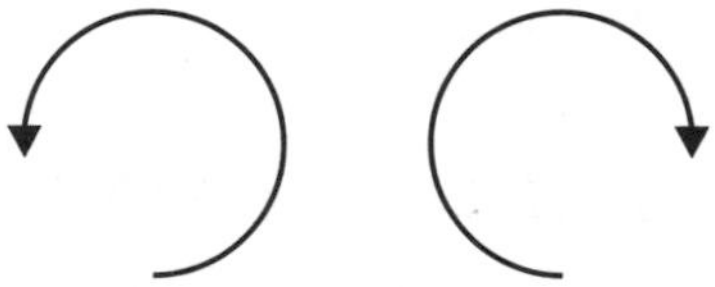

When calculating a moment a force is multiplied by a distance, so the SI units of a moment are Newton metres or N m.

Worked example 1

Find the moment of each of these forces about the point O.

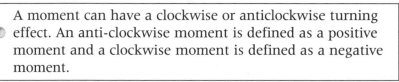

(a)

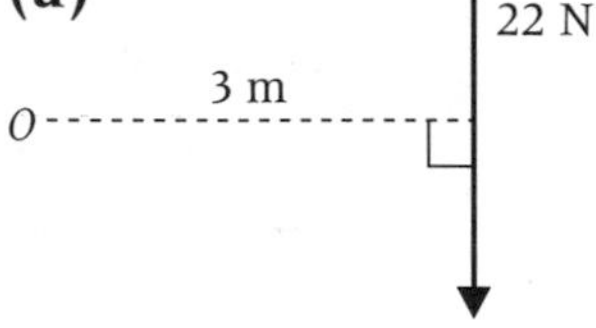

(b)

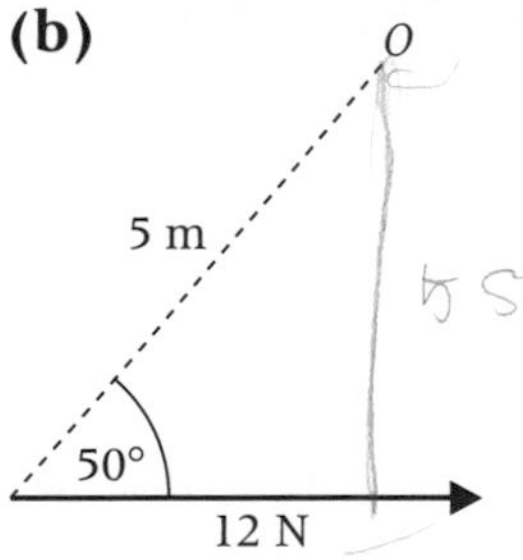

(c)

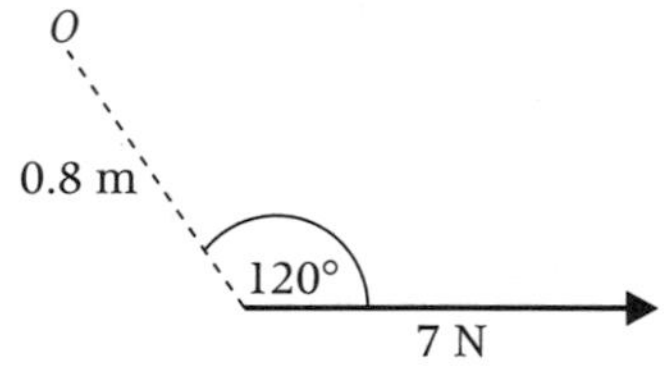

Solution

(a) Here the perpendicular distance is 3 m.

$$\text{Moment about } O = -22 \times 3 = -66 \text{ N m}$$

The force produces a clockwise moment and so is negative.

(b) Here the perpendicular distance must be drawn in, as shown in the diagram. In this case it is 5 sin 50°.

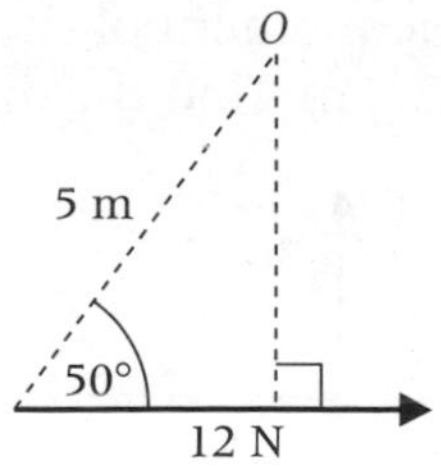

$$\text{Moment about } O = 12 \times 5 \sin 50°$$
$$= 46.0 \text{ N m}$$

The moment is anti-clockwise and so is positive.

(c) To find the perpendicular distance in this case the line of action of the force must be drawn, by extending the arrow used to represent the force. The perpendicular distance can then be calculated as 0.8 sin 60°.

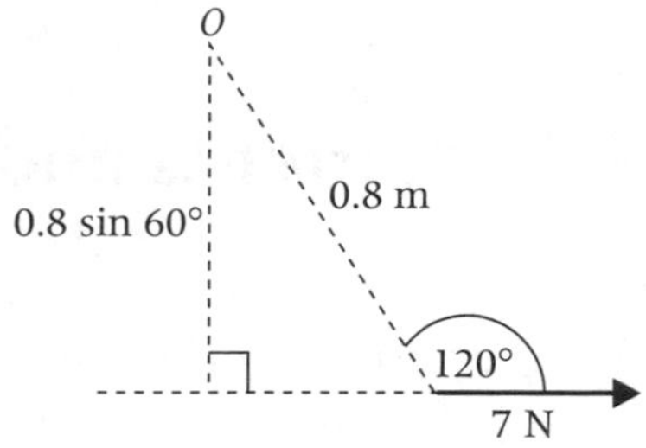

$$\text{Moment about } O = 7 \times 0.8 \sin 60°$$
$$= 4.85 \text{ N m}$$

The moment is anti-clockwise and so is positive.

EXERCISE 9A

1 Find the moment of each of the forces below about the point O.

(a)

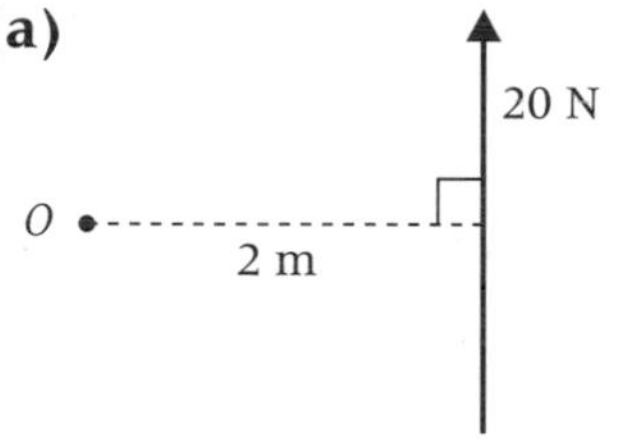

(b)

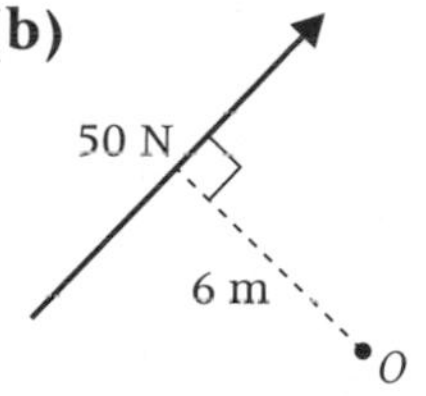

(c)

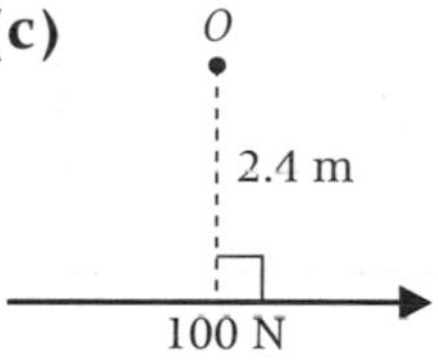

(d)

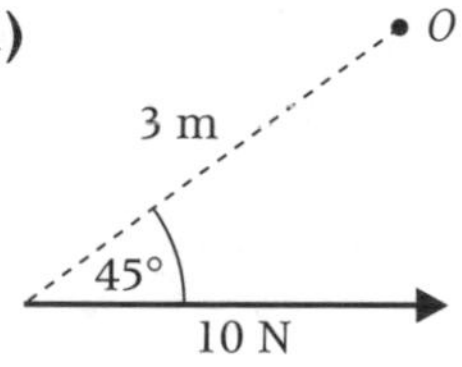

(e)

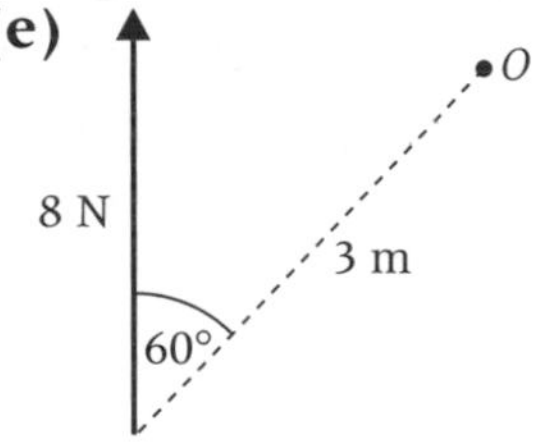

(f)

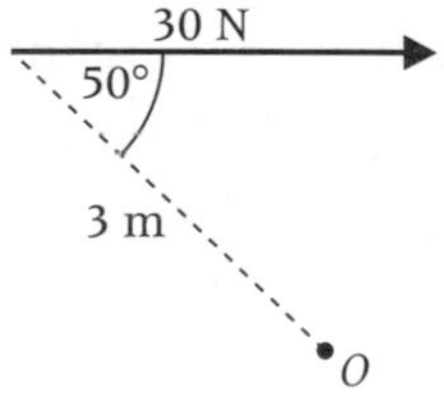

(g)

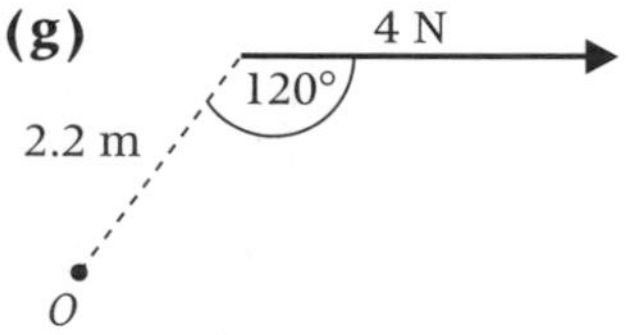

(h)

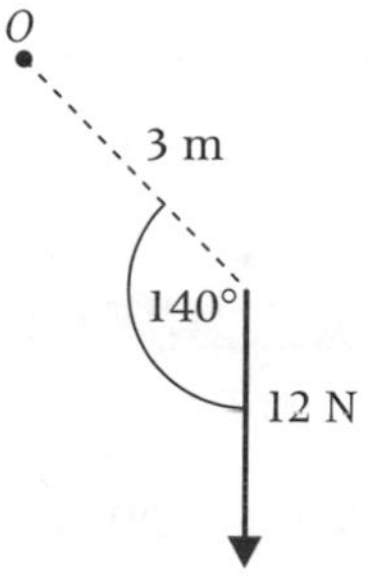

(i)

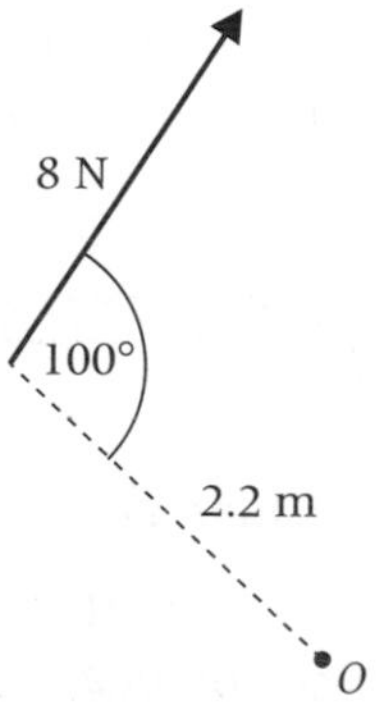

2 If the moment of each force below about the point O is -40 N m, find the distance d in each case.

(a)

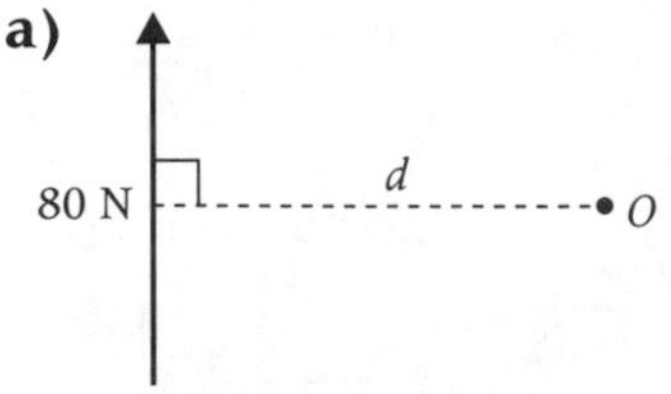

(b)

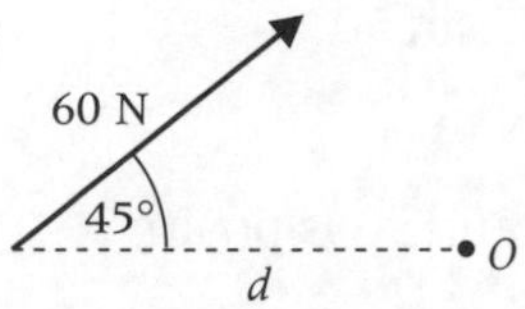

(c)

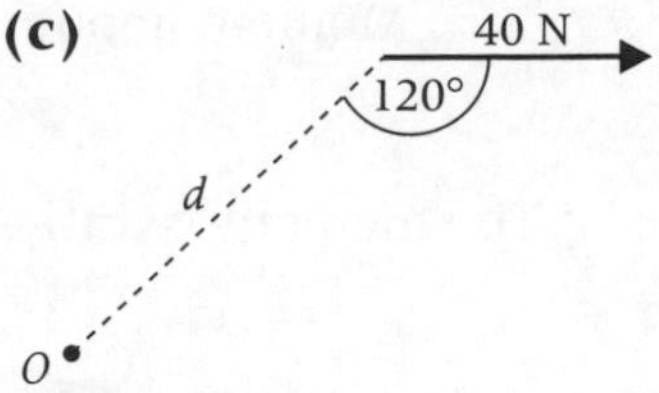

9.3 Moments and equilibrium

> For a particle to be in equilibrium, the resultant force on the particle must be zero. For a rigid body to be in equilibrium the resultant force must be zero and the total moment of all the forces acting must also be zero.

This is illustrated in the diagram below.

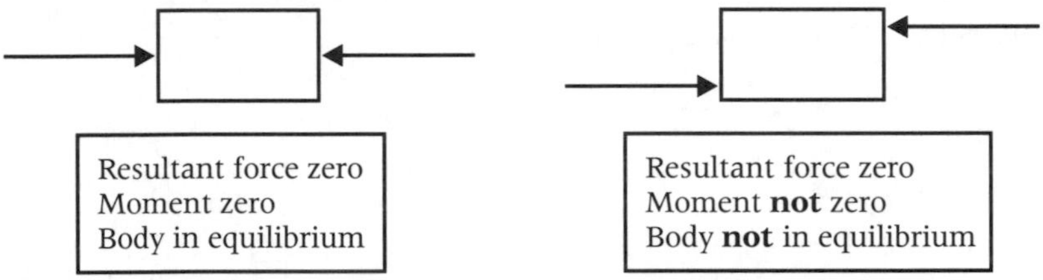

The following examples show how this principle can be applied to physical situations.

In all the examples below we will consider uniform bodies. These are bodies that are made from the same material throughout. These bodies will also be symmetric shapes so that we can assume that the force of gravity or weight acts at the centre of the object.

Worked example 2

A uniformed beam of mass 20 kg and length 3 m rests on two supports as shown below.

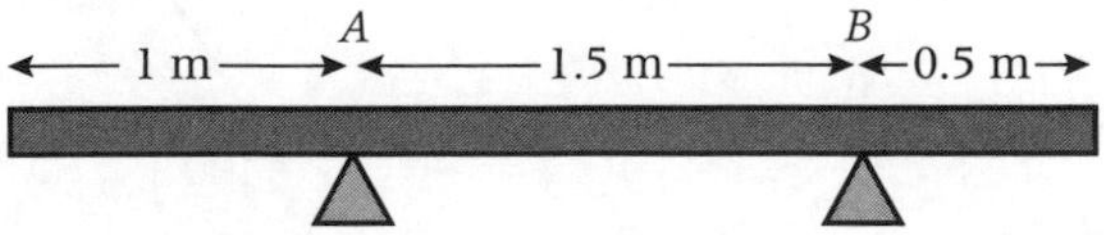

(a) Find the reaction force exerted by each support.

(b) Find the greatest mass that can be placed at the left-hand end of the beam.

Solution

(a) The diagram shows the forces acting on the beam.

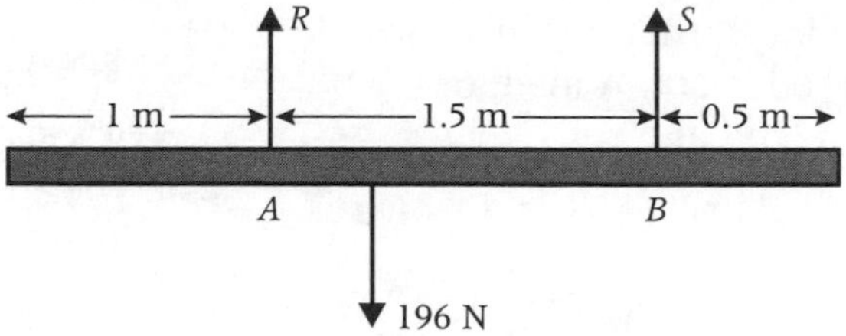

As the upward forces must balance the downward force.

$$R + S = 196$$

Taking moments about the point A gives:

$$0.5 \times 196 = 1.5S$$

$$S = \frac{98}{1.5}$$

$$= 65\frac{1}{3}\text{ N}$$

Then using $R + S = 196$ gives:

$$R + 65\frac{1}{3} = 196$$

$$R = 130\frac{2}{3}\text{ N}$$

(b) If a mass is placed at the left hand end of the beam, then an extra force is added to the diagram as shown.

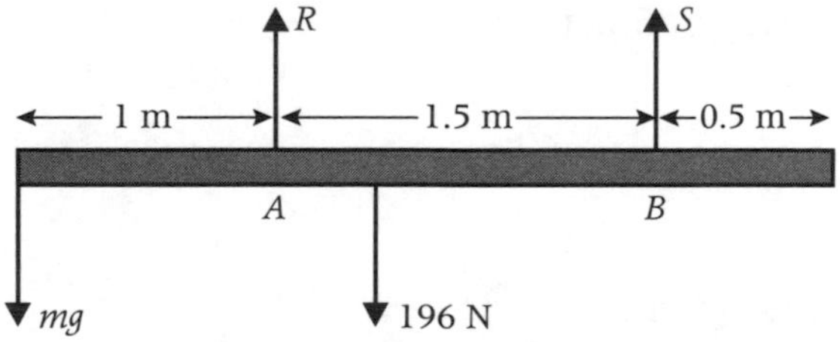

When the beam is just in equilibrium with the maximum possible mass, the reaction force, S, will be zero. The beam is effectively balanced on the support at A.

Taking moments about A and with $S = 0$ gives:

$$1 \times 9.8m = 0.5 \times 196$$

$$m = 10\text{ kg}$$

Worked example 3

The diagram shows rod, of mass 30 kg and length 3 m, that is smoothly hinged at A. The rod is held in a horizontal position by a rope. The rope is attached to the rod at a point B, that is 2 m from A. The angle between the rope and the rod is 60°. A load, of mass 100 kg, is suspended from the end of the rod at C.

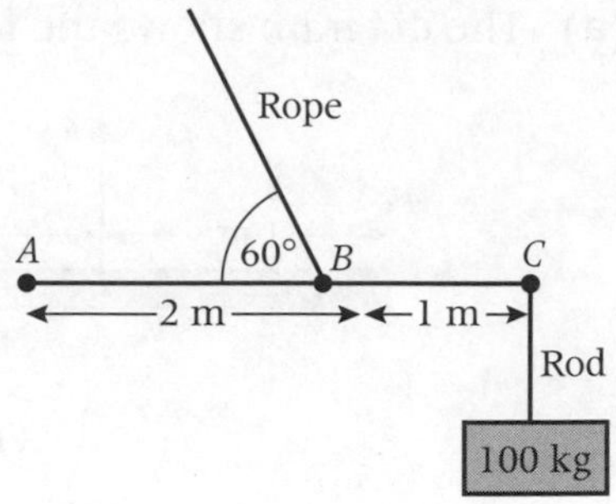

Find the tension in the rope.

Solution

The diagram shows the forces acting. We will find the tension by taking moments about the point A.

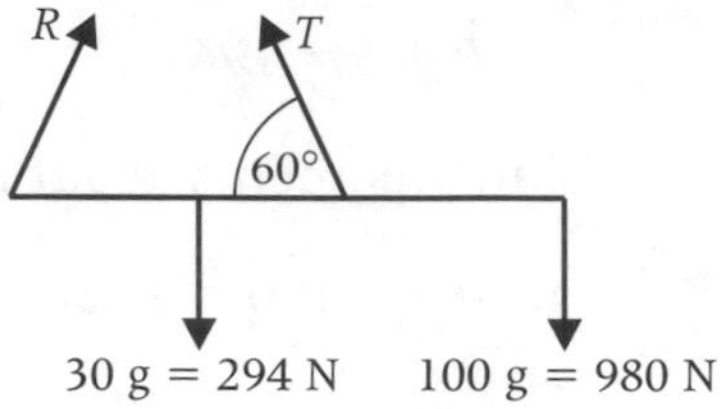

The perpendicular from A to the tension force is 2 sin 60°.

Taking moments about A gives

$$T \times 2 \sin 60° = 100 \times 9.8 \times 3 + 30 \times 9.8 \times 1.5$$

$$T = \frac{3381}{2 \sin 60°}$$

$$= 1952 \text{ N}$$

Note that because we take moments about A, we do not need to worry about the size of the reaction force, R, that acts at A.

Worked example 4

A ladder, of length 5 m and mass 20 kg, leans against a smooth wall so that it is at an angle of 60° to the horizontal. The ladder remains at rest, with its base on rough, horizontal ground.

(a) Find the magnitude of the normal reaction and friction forces acting on the base of the ladder.

(b) Find an inequality that the coefficient of friction must satisfy.

Solution

(a) The diagram shows the forces acting on the ladder.

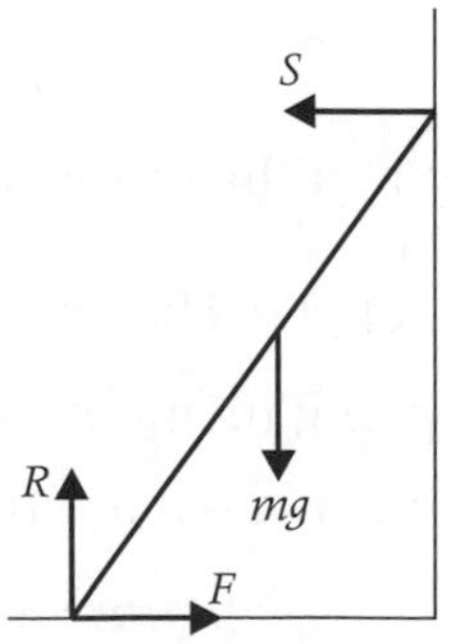

As the vertical forces acting on the ladder are in equilibrium, we have:

$$R = 20g$$

$$= 196 \text{ N}$$

As the horizontal forces are also in equilibrium, we have:

$$F = S$$

Next we will take moments about the base of the ladder. For the weight, mg, the perpendicular distance from the base to the force is 2.5 cos 60° and the perpendicular distance from the base to the force S is 5 cos 30°. Hence taking moments about the base gives:

$$S \times 5 \cos 30° - 196 \times 2.5 \cos 60° = 0$$

$$S = \frac{196 \times 2.5 \cos 60°}{5 \cos 30°}$$

$$= 56.58 \text{ N}$$

But as $F = S$, we have $F = 56.58$ N.

(b) As the ladder is at rest the friction inequality $F \leqslant \mu R$ must be satisfied. Substituting for R and F gives:

$$F \leqslant \mu R$$

$$56.58 \leqslant \mu \times 196$$

$$\mu \geqslant \frac{56.58}{196} = 0.289 \text{ (to three significant figures)}$$

Worked example 5

A boom on a yacht is held in a horizontal position by a rope attached to the top of the mast. The boom is freely pivoted where it is attached to the mast. The length of the boom is 4 m and its mass is 15 kg. The rope makes an angle of 70° with the boom.

(a) Find the tension in the rope.

(b) Find the magnitude of the force that the mast exerts on the boom.

9

Solution

(a) The diagram shows the forces acting on the boom.

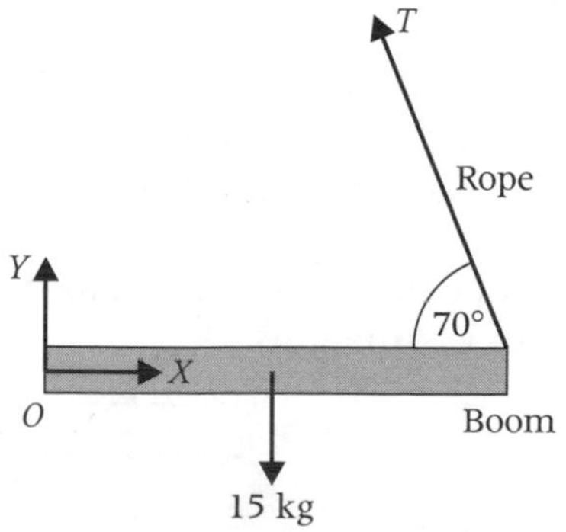

Taking moments about the point, O, where the boom is attached to the mast gives:

$$T \times 4 \sin 70° = 15 \times 9.8 \times 2$$

$$T = \frac{294}{4 \sin 70°}$$

$$= 78.2 \text{ N}$$

(b) To find the force exerted by the mast, assume that it has two components, X horizontal and Y vertical.

Then for horizontal equilibrium, we require:

$$\begin{aligned} X &= T\cos 70° \\ &= \frac{294\cos 70°}{4\sin 70°} \\ &= 26.8\text{ N} \end{aligned}$$

For vertical equilibrium

$$Y + T\sin 70° = 15 \times 9.8$$

$$Y = 147 - \frac{294}{4}$$

$$= 73.5\text{ N}$$

The magnitude of the force acting is given by $\sqrt{X^2 + Y^2}$.

Substituting the values obtained above gives 78.2 N.

EXERCISE 9B

1 A see-saw of length 4 m, pivoted at its centre, rests in a horizontal position. John, who has mass 30 kg, sits at one end. Where should his friend James, who has mass 40 kg, sit if the see-saw is to remain in a horizontal position?

2 A uniform beam of mass 30 kg and length 8 m, rests on supports that are 2 m from each end. A load of mass 50 kg is suspended from a point 3 m from one end. Find the magnitude of the reaction forces exerted by the two supports.

3 The diagram shows a metal beam of mass 40 kg that rests on two supports.

(a) Find the magnitude of the two reaction forces acting on the beam.

(b) A load of mass 30 kg is suspended from the beam, at a point 4 m from one end. Find the reaction forces at each end of the beam.

4 A beam of length 3 m and mass 15 kg is supported as shown in the diagram.

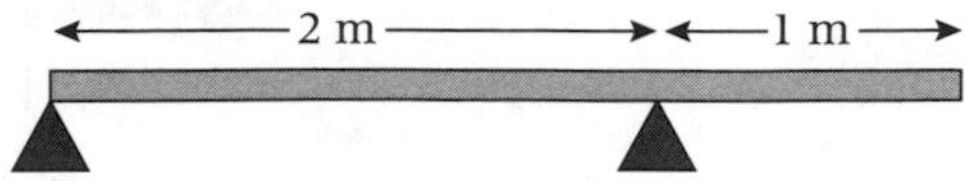

(a) Find the magnitude of each of the reaction forces acting on the beam.

(b) A mass is placed at the unsupported end of the beam. What is the greatest mass that can be placed in this position?

5 The diagram shows a uniform beam of mass 20 kg that rests on two concrete blocks.

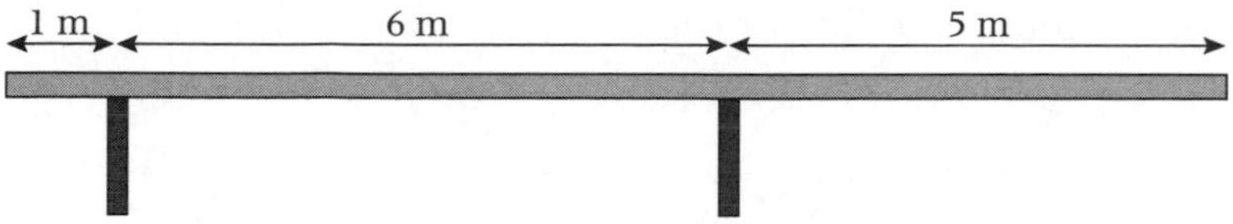

(a) Find the magnitude of each of the reaction forces acting on the beam.

A box is to be placed on the beam.

(b) What is the greatest mass of a box that could be placed at either end of the beam?

6 A plank of length 1.6 m and mass 4 kg rests on two supports which are 0.3 m from each end of the plank. A mass is attached to one end of the plank. If the normal reaction force on the support nearer to this load is twice the normal reaction force on the other support, determine the mass attached.

7 A plank of length 2 m and mass 6 kg is suspended in a horizontal position by two vertical strings, one at each end. A 2 kg mass is placed on the plank at a variable point P. If either string snaps when the tension in it exceeds 42 N, find the section of the plank in which P can be.

8 A metre rule is pivoted 20 cm from one end A and is balanced in a horizontal position by hanging a mass of 180 g at A. What is the mass of the rule? What additional mass should be hung from A if the pivot is moved 10 cm nearer to A?

9 A metre rule of mass 100 g is placed on the edge of a table as shown with a 200 g mass at A. A mass M g is attached at C.

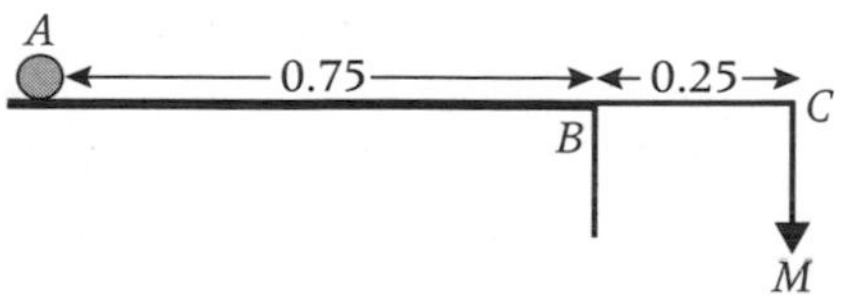

(a) When the rule is just on the point of overturning, where does the normal contact force act?

(b) Determine the maximum value of M for which the rule will not overturn.

9

10 The diagram shows the forces that act on a rectangular sheet of metal. All the forces lie in the same plane as the sheet.

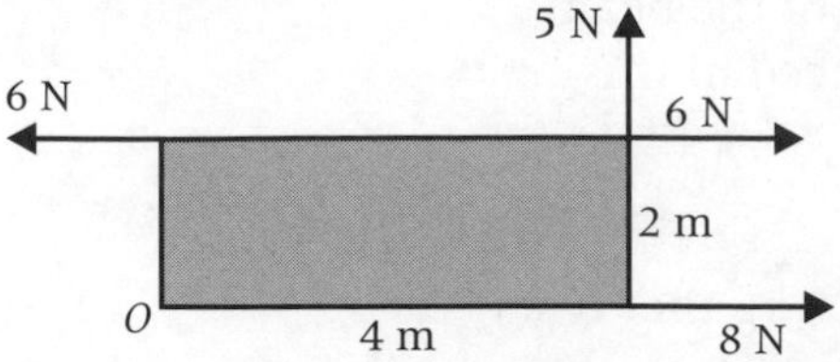

(a) Find the resultant moment of the four forces about the corner O.

(b) A force of magnitude 10 N is applied to the sheet, so that the total moment about O is zero. Draw a diagram to show where this force should be applied.

11 The diagram shows a light rod, of length 2 m, that is smoothly pivoted at O. A horizontal force of 200 N acts at B, which is at the centre of the rod. Find the magnitude of the force that acts at A if the rod is in equilibrium.

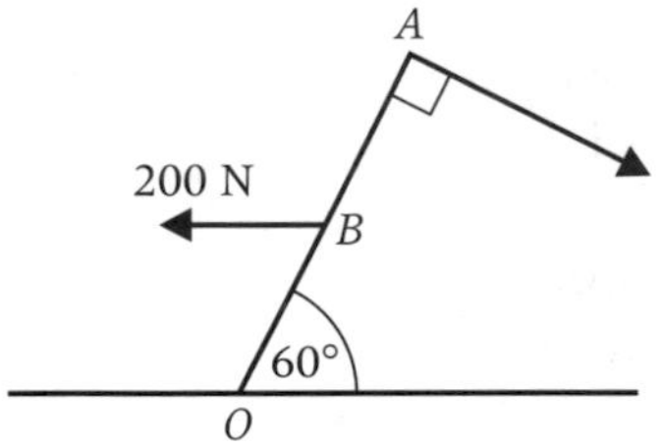

12 A uniform plank of length 2 m and mass 5 kg is connected to a vertical wall by a smooth hinge at A and a wire CB as shown. If a 10 kg mass is attached to D, find:

(a) the tension in the wire

(b) the magnitude of the reaction at the hinge A.

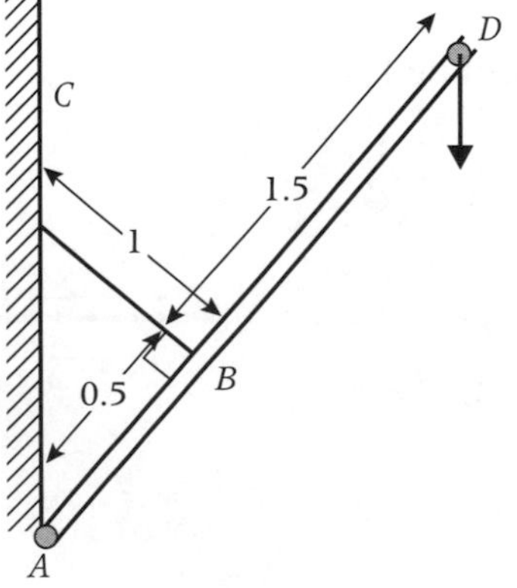

13 A loft door OA of weight 100 N is propped open at 50° to the horizontal by a strut AB. The door is hinged at O; $OA = OB = 1$ m. Assuming that the mass of the strut can be neglected compared to the mass of the door and that the weight of the door acts through the midpoint of OA, find:

(a) the force in the strut

(b) the reaction at the hinge.

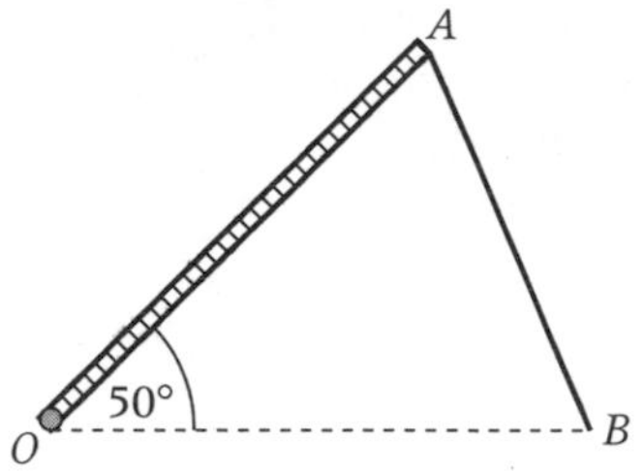

14 A ladder, of length 3 m and mass 20 kg, leans against a smooth, vertical wall so that the angle between the horizontal ground and the ladder is 60°. Find the magnitude of the friction and normal reaction forces that act on the ladder, if it is in equilibrium.

15 The foot of a uniform ladder of mass m rests on rough horizontal ground and the top of the ladder rests against a smooth vertical wall. When a man of mass $4m$ stands at the top of the ladder the system is in equilibrium with the ladder inclined at 60° to the horizontal. Show that the coefficient of friction between the ladder and the ground is greater than or equal to $\frac{3\sqrt{3}}{10}$. [A]

16 A uniform ladder, of mass m, leans against a vertical wall with its base on horizontal ground. The length of the ladder is 6 m. Assume that the wall is smooth and that the ground is rough, with the coefficient of friction between the ladder and the ground equal to 0.5.

(a) If the angle between the ladder and the ground is θ, show that the ladder remains at rest if θ is greater than or equal to 45°.

A person of mass M climbs the ladder.

(b) Show that when the person is at a distance x m from the bottom of the ladder

$$\tan\theta \geqslant \frac{3m + Mx}{3(m + M)}$$

if the ladder is to remain at rest.

9

(c) How far up the ladder can the person climb if $\theta = 45°$?

(d) Now assume that $\tan \theta = 2$.

(i) Show that $x \leqslant 6 + \dfrac{3m}{M}$, if the ladder remains at rest.

(ii) Use this result to make a prediction about the mass of a person who can reach the top of the ladder. [A]

17 The diagram shows a man of mass 70 kg at rest while abseiling down a vertical cliff. Assume that the rope is attached to the man at his centre of mass. *You should model the man as a uniform rod and assume that he is not holding the rope.*

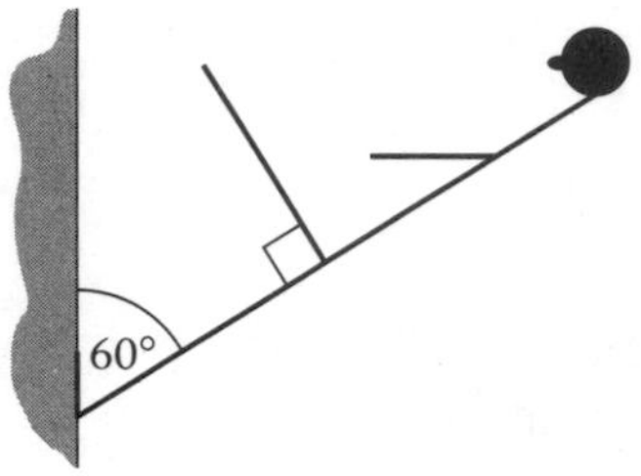

(a) Draw a diagram to show the forces acting on the man.

(b) The angle between the man's legs and the cliff is 60°. By taking moments show that the man can only remain in this position if the coefficient of friction between his feet and the wall is greater than or equal to $\dfrac{1}{\sqrt{3}}$.

(c) Find the tension in the rope.

(d) In the position shown above, the rope is at 90° to the man's body. Find the magnitude of the reaction force on the man's feet. [A]

18 The diagram shows a man holding a rope attached to one end of a uniform metal rod, of mass 200 kg, that is freely pivoted at A. A 120 kg mass is attached to the other end of the rod by another rope. Initially the rod is horizontal.

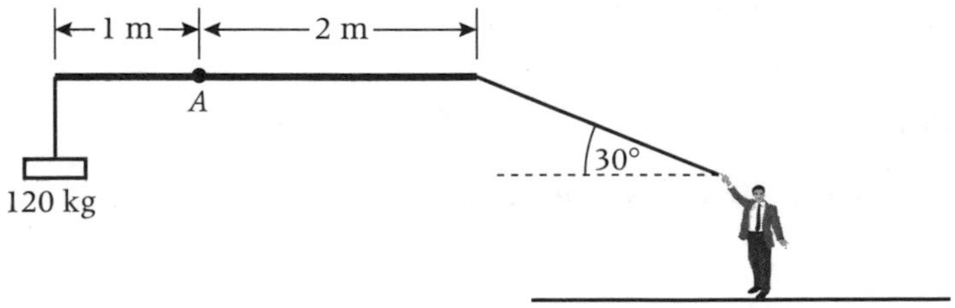

(a) Draw a labelled diagram to show the forces acting on the rod.

(b) Find the tension in the rope that the man is holding, when the rod is horizontal.

(c) The mass of the man is 70 kg and he is modelled as a particle. Using the tension found in part **(b)**, show that he remains at rest if the coefficient of friction between the ground and his feet is greater than or equal to $\frac{\sqrt{3}}{6}$.

State, with a reason, whether or not the man is likely to remain at rest.

(d) The man tries to pull harder so that the rope is in line with the rod as shown below.

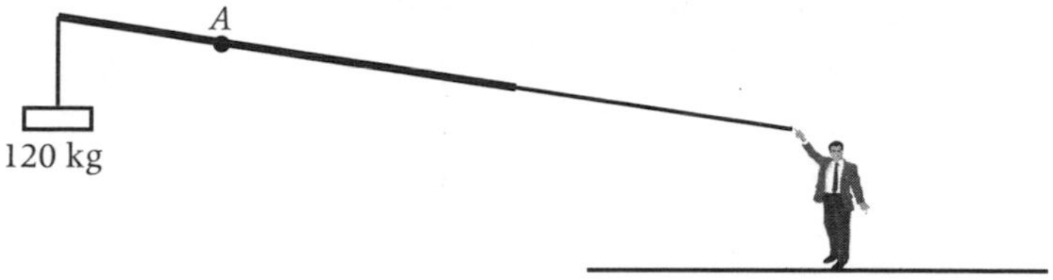

Explain whether this is possible. [A]

9.4 Centre of mass of a system of particles

In this section we will consider how to find the centre of mass of a system of particles. To illustrate the principle that we will use, imagine a light rod which has a different sized mass fixed to each end. There will be a point on the rod at which it can be balanced. This point is called the centre of mass.

The position of the centre of mass can be found using moments.

The diagram shows a rod with two particles of masses, m_1 and m_2, which are at distances x_1 and x_2 from the left-hand end. Assume that the rod can be balanced by a single force, R, acting upwards at a distance $\bar{x}$ from the left-hand end.

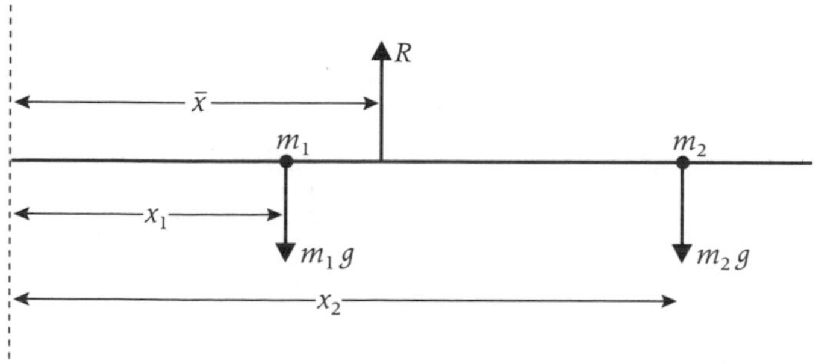

For the rod to balance the upward force must balance the two downwards forces, so that $R = m_1 g + m_2 g$.

Now taking moments about the left-hand end of the rod gives:

$$R\bar{x} - m_1 g x_1 - m_2 g x_2 = 0$$

$$\bar{x} = \frac{m_1 g x_1 + m_2 g x_2}{R}$$

$$= \frac{m_1 g x_1 + m_2 g x_2}{m_1 g + m_2 g}$$

$$= \frac{m_1 x_1 + m_2 x_2}{m_1 + m_2}$$

This principle can be extended to find the centre of mass of any number of particles, using the general result stated below.

$$\bar{x} = \frac{\sum_{i=1}^{n} m_i x_i}{\sum_{i=1}^{n} m_i}$$

The worked examples below illustrate how this can be applied.

Worked example 6

Particles of masses 5 kg, 3 kg and 2 kg are fixed to a light rod of length 1.2 m. The 3 kg mass is in the centre and the others are the ends of the rod. Find the distance of the centre of mass from the 5 kg mass.

Solution

The diagram shows the positions of the masses on the rod.

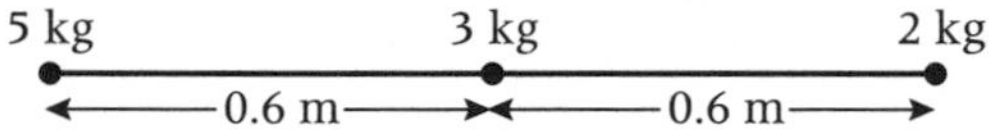

Using the formula $\bar{x} = \dfrac{\sum_{i=1}^{n} m_i x_i}{\sum_{i=1}^{n} m_i}$

$$\bar{x} = \frac{5 \times 0 + 3 \times 0.6 + 2 \times 1.2}{5 + 3 + 2}$$

$$= \frac{4.2}{10}$$

$$= 0.42 \text{ m}$$

Working in two dimensions

For a system of particles in two dimensions the result above can still be used. The diagram shows a system of four particles in two dimensions.

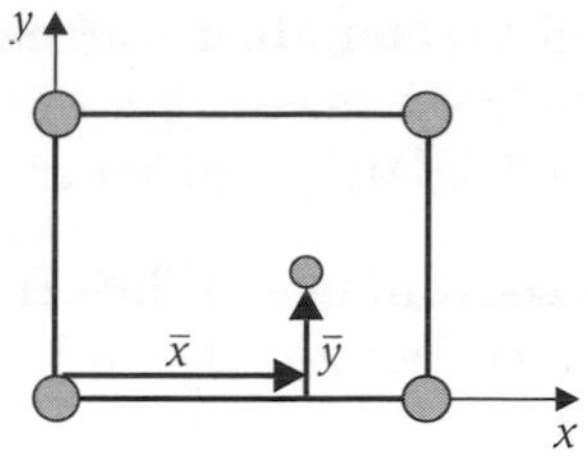

The position of the centre of mass relative to the bottom left hand corner of the system has been shown. The coordinates of this point relative to the corner are $\bar{x}$ and $\bar{y}$. These are calculated using

$$\bar{x} = \frac{\sum_{i=1}^{n} m_i x_i}{\sum_{i=1}^{n} m_i} \text{ and } \bar{y} = \frac{\sum_{i=1}^{n} m_i y_i}{\sum_{i=1}^{n} m_i}$$

where x_i and y_i are the coordinates of the particles.

Worked example 7

Particles of mass 4 kg, 5 kg, 8 kg and 2 kg are fixed to the corners of a light square framework, with sides of length 0.6 m. The diagram shows the positions of the masses.

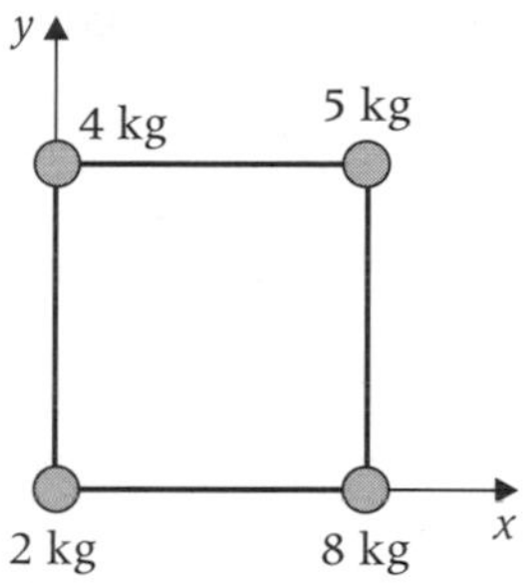

Find the coordinates of the centre of mass of the system, with respect to the 2 kg mass.

Solution

Using the formula $\bar{x} = \frac{\sum_{i=1}^{n} m_i x_i}{\sum_{i=1}^{n} m_i}$ gives:

$$\bar{x} = \frac{0 \times 2 + 0 \times 4 + 0.6 \times 8 + 0.6 \times 5}{2 + 4 + 8 + 5}$$

$$= \frac{7.8}{19}$$

$$= 0.41$$

Similarly using the formula $\bar{y} = \frac{\sum_{i=1}^{n} m_i y_i}{\sum_{i=1}^{n} m_i}$ gives:

$$\bar{y} = \frac{0 \times 2 + 0.6 \times 4 + 0 \times 8 + 0.6 \times 5}{2 + 4 + 8 + 5}$$

$$= \frac{5.4}{19}$$

$$= 0.28$$

So the coordinates relative to the 2 kg mass are (0.41, 0.28), where the x axis is horizontal and the y axis is vertical.

9

Worked example 8

A light rectangular framework has sides of length 2 m and 1 m. Particles of mass 2 kg, 4 kg, M kg and m kg are fixed to the corners of the framework as shown in the diagram.

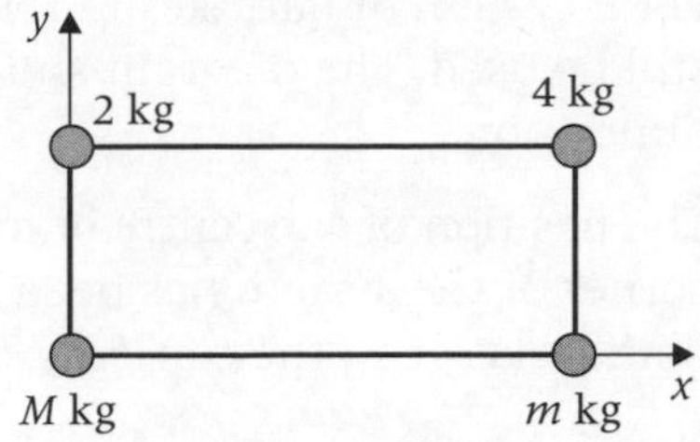

The coordinates of the centre of mass relative to the bottom left hand corner are (1.6, 0.3). Find M and m.

Solution

First consider the x coordinate of the centre of mass.

$$1.6 = \frac{0 \times M + 0 \times 2 + 2 \times 4 + 2 \times m}{2 + 4 + m + M}$$

$$1.6(6 + m + M) = 8 + 2m$$

$$1.6 = 0.4m - 1.6M$$

$$4 = m - 4M \qquad (1)$$

Now consider the y coordinate:

$$0.3 = \frac{0 \times M + 1 \times 2 + 1 \times 4 + 0 \times m}{2 + 4 + m + M}$$

$$0.3(6 + m + M) = 6$$

$$4.2 = 0.3m + 0.3M$$

$$14 = m + M \qquad (2)$$

We now have a pair of simultaneous equations. Eliminating m from these equations, by subtracting (1) from (2), gives:

$$10 = 5M$$

$$M = 2\text{ kg}$$

Then substituting this value into equation (2) gives:

$$14 = m + 2$$

$$m = 12\text{ kg}$$

Worked example 9

A rod has mass 5 kg and length 60 cm. A particle of mass 12 kg is fixed at one end and a particle of mass 8 kg is fixed at the other. Find the distance of the centre of mass from the 12 kg mass.

Solution

Model the rod as a particle of mass 5 kg at its centre, as shown in the diagram.

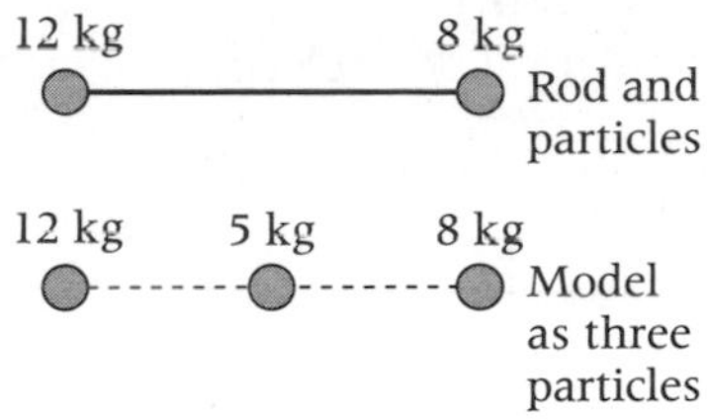

Then using the formula $\bar{x} = \frac{\sum_{i=1}^{n} m_i x_i}{\sum_{i=1}^{n} m_i}$ gives:

$$\bar{x} = \frac{12 \times 0 + 5 \times 0.3 + 8 \times 0.6}{12 + 5 + 8}$$

$$= \frac{6.3}{25}$$

$$= 0.252 \text{ m or } 25.2 \text{ cm}$$

EXERCISE 9C

1 Particles of mass 3 kg and 6 kg are placed at opposite ends of a light rod of length 1.8 m. Find the distance of the centre of mass from the 3 kg mass.

2 A 5 kg particle is placed at the centre of a light rod of length 0.8 m. A particle of mass 2 kg is fixed to one end of the rod and a particle of mass 1 kg is fixed to the other end. Find the distance of the centre of mass from the 1 kg mass.

3 Four particles of mass 5 kg, 4 kg, 3 kg and 6 kg lie on the same horizontal line as shown. How far is their centre of mass from the 5 N particle?

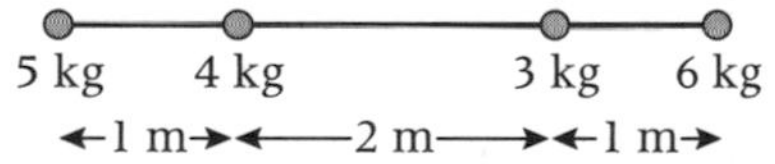

4 Three particles A, B, C, of masses 5 kg, 3 kg, 4 kg, respectively, lie on the same horizontal line as shown. Find the distance of their centre of mass from A.

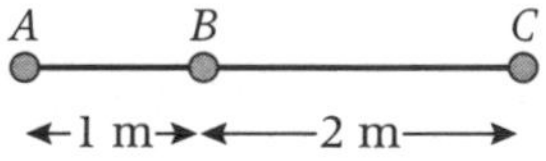

The particle A is removed and replaced by a new particle. What is the greatest value of its mass if the centre of mass of the three particles is to lie in BC?

5 A rod, of length 50 cm, has three masses attached to it. There is an 8 kg mass at one end and a 7 kg mass at the other end. A 5 kg mass is also attached at another point on the rod. The centre of mass of the rod is 20 cm from the 8 kg mass.

Find how far the 5 kg mass is from the 8 kg mass, if:

(a) the rod is light

(b) the rod is uniform and has a mass of 5 kg.

6 For each light framework shown below find the coordinates of the centre of mass of the particles attached to the framework, with respect to the corner marked O.

(a)

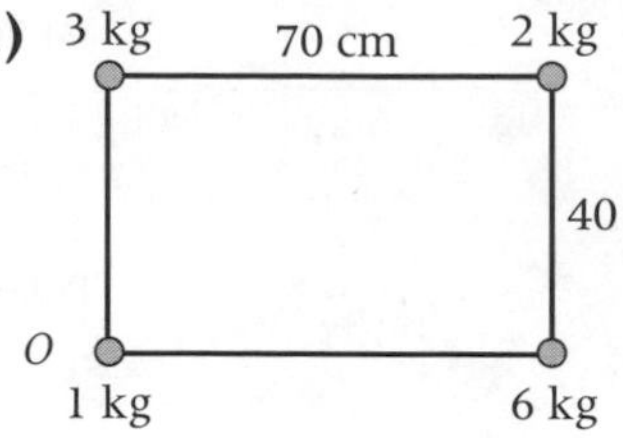

(b)

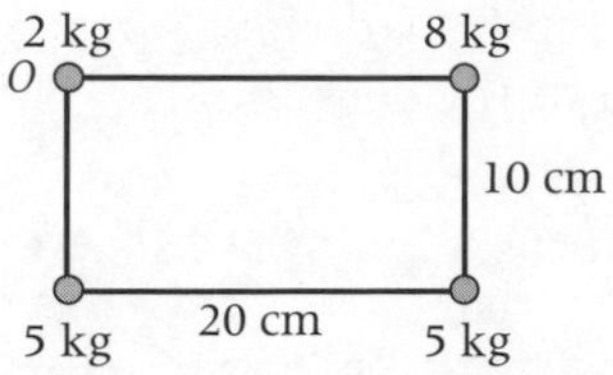

(c)

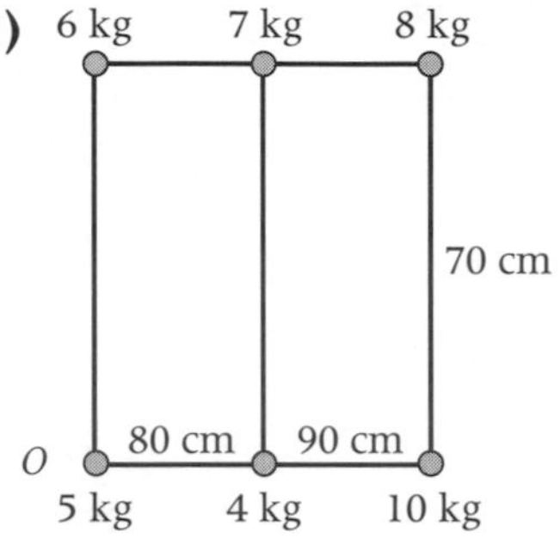

(d)

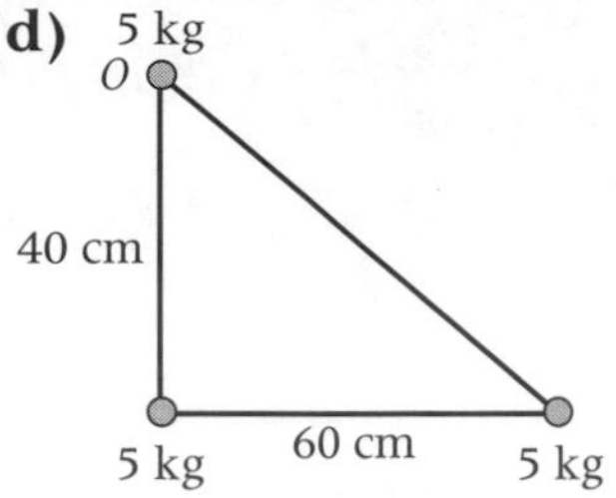

7 Particles of mass m, $2m$, $3m$ and $4m$ are attached to the corners of a square framework with sides of length l. The framework is shown in the diagram. Find the position of the centre of mass with respect of the corner marked O, if:

(a) the framework is light

(b) the framework is made up of 4 rods of mass m.

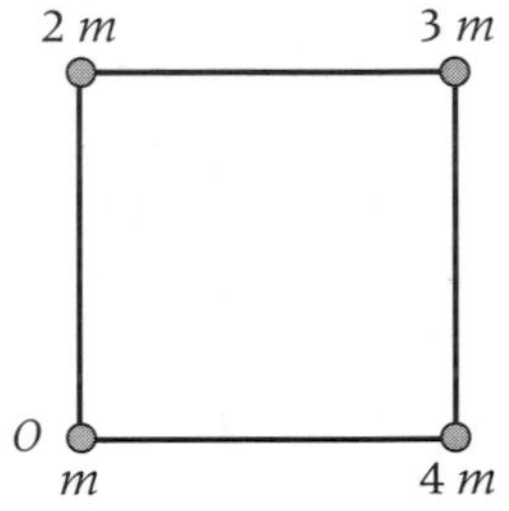

8 The diagram shows a light rectangular framework of height 40 cm and width 20 cm. Particles are attached to the corners of the framework as shown in the diagram. The centre of mass is at a height of 16 cm and at a distance of 14 cm from the left-hand edge of the framework.

Find m and M.

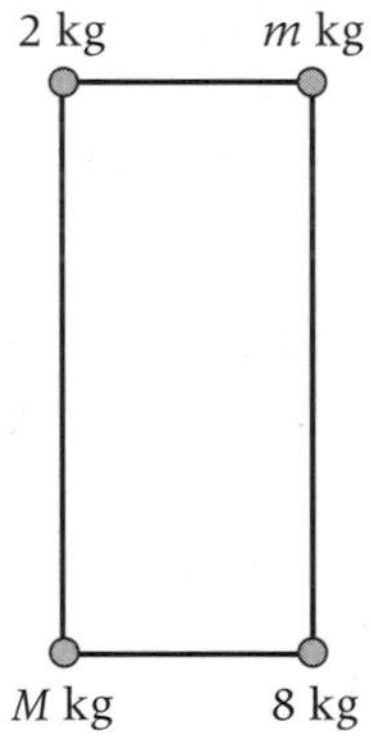

9 A light, triangular framework has sides of length 30 cm, 40 cm and 50 cm. Particles of mass 3 kg, 2 kg and m kg are attached to it as shown in the diagram.

(a) Find m if the centre of mass is to be at a height of 10 cm above the base.

(b) Find the distance of the centre of mass from the 3 kg mass.

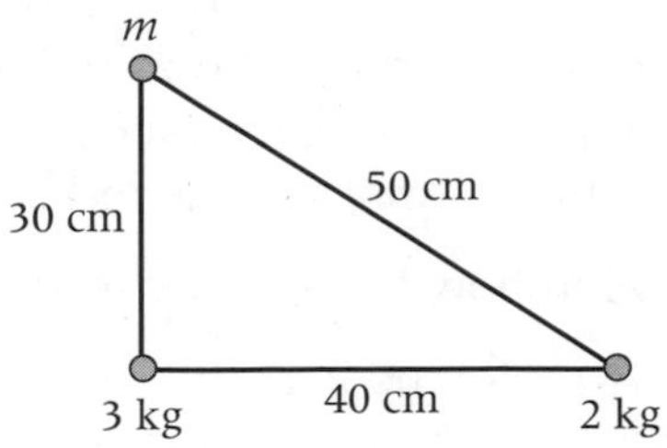

10 Particles are fixed to the ends of a light cross shape as shown in the diagram. The cross is made from two light rods of length 40 cm. The rods are joined at their centres, so that they are perpendicular. Find the distance of the centre of mass of the system from the centre of the cross, O.

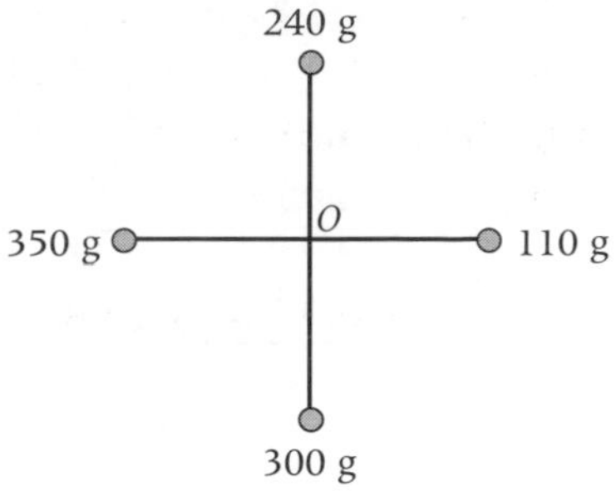

9.5 Centre of mass of a composite body

In this section we will find the centres of mass of composite bodies. By a composite body we mean something formed when two or more parts are joined together, for example a disc attached to the end of a rod. We will also consider shapes that have had holes cut in them. Some simple composite bodies are illustrated.

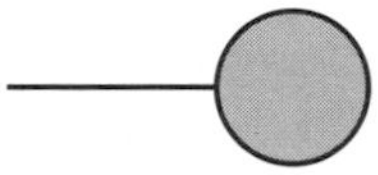

A disc attached to a rod

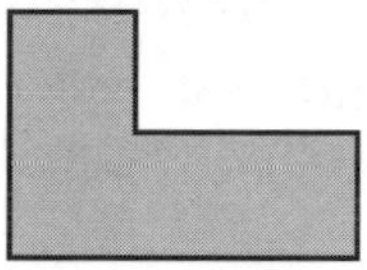

A body formed by joining two rectangles

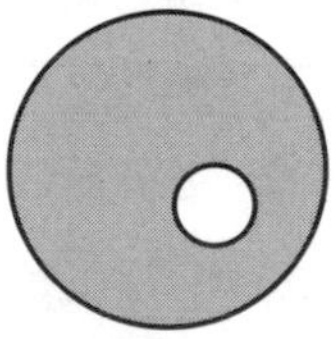

A circle with a hole cut in it

To find the centre of mass of a composite body we break it down into two or more shapes and find the centre of mass of each of these shapes. This is very often obvious because the shapes will be simple ones like rectangles, squares or circles. Then the centre of mass can be found by using the same approach that we have used for particles in the last section, by assuming that there is a particle at the centre of mass of each of the shapes that are being considered.

You will find the word lamina is often used in this context. A lamina is a thin sheet of material. The thickness of the lamina is negligible. Also when working with composite body problems it is important to check that the bodies are uniform, that is they are made entirely from the same material.

9

Worked example 10

A disc of mass 4 kg and radius 18 cm is attached to the end of a rod of mass 5 kg and length 180 cm. Find the distance of the centre of mass from the base of the rod.

Solution

The diagram shows the centres of mass of the rod and the disc.

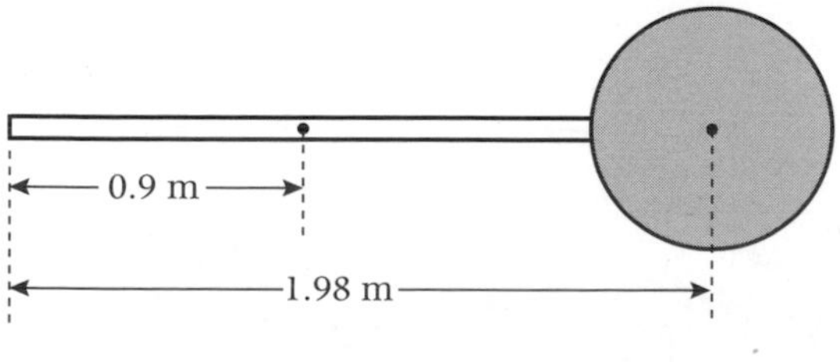

Using the formula $\bar{x} = \dfrac{\sum_{i=1}^{n} m_i x_i}{\sum_{i=1}^{n} m_i}$ gives:

$$\bar{x} = \frac{5 \times 0.9 + 4 \times 1.98}{4 + 5}$$

$$= \frac{12.42}{9}$$

$$= 1.38 \text{ m}$$

Centre of mass of a lamina

When working with a uniform lamina, as in the examples below, it is possible to work with areas instead of masses, because the area of each part will be proportional to its mass.

Worked example 11

The diagram shows a uniform lamina. Find the distance of the centre of mass from *AB* and the distance from *AF*.

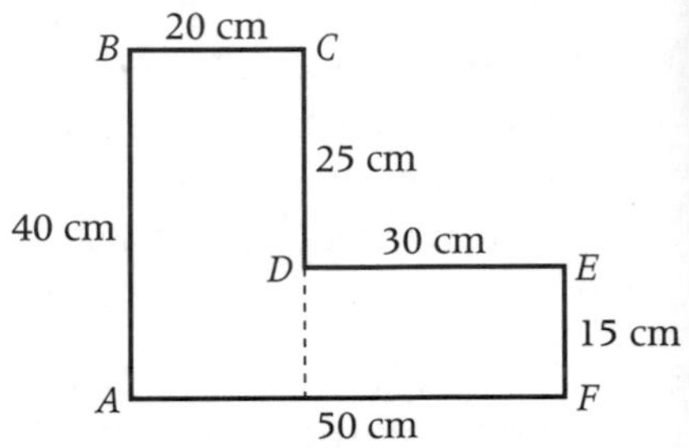

Solution

First find the total area of the lamina, which has been split into two parts on the diagram.

$$\text{Area} = 40 \times 20 + 30 \times 15$$

$$= 1250 \text{ cm}^2$$

To find the distance of the centre of mass from *AB* use the formula $\bar{x} = \dfrac{\sum_{i=1}^{n} m_i x_i}{\sum_{i=1}^{n} m_i}$, but replacing the masses with the areas of each part.

$$\bar{x} = \frac{800 \times 10 + 450 \times 35}{1250}$$

$$= 19 \text{ cm}$$

Similarly to find the distance from *AF*, use the formula

$$\bar{y} = \frac{\sum_{i=1}^{n} m_i y_i}{\sum_{i=1}^{n} m_i}$$

$$\bar{y} = \frac{800 \times 20 + 450 \times 7.5}{1250}$$

$$= 15.5 \text{ cm}$$

The position of a suspended lamina or body

When a lamina or other body is suspended, it will hang, so that the centre of mass is directly below the point of suspension.

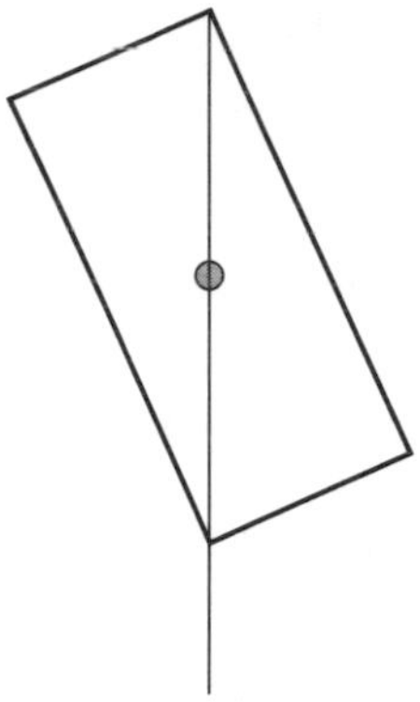

If you take a rectangle and suspend it from one corner, the rectangle will hang so that one diagonal is vertical, because the central of mass is on the diagonal, as shown in the diagram.

The angle between the vertical and one side in this position can then be calculated.

Worked example 12

The lamina from the last example is suspended from the corner B. The lamina remains at rest. Find the angle between the side AB and the vertical.

Solution

The diagram shows the lamina, as it would hang, with the centre of mass directly below B.

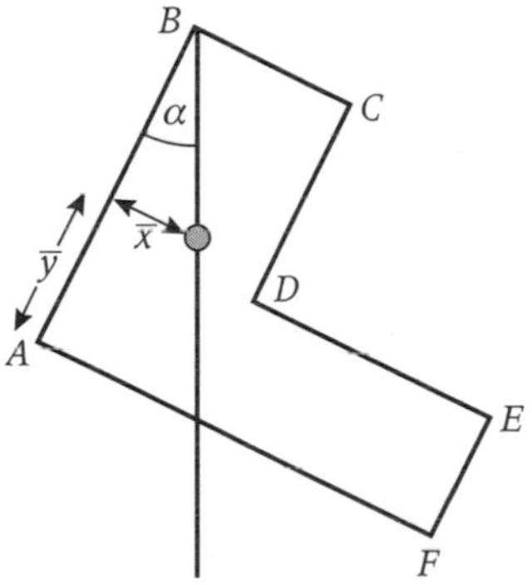

To find the angle α, as required, note that:

$$\tan \alpha = \frac{\overline{x}}{40 - \overline{y}}$$

$$= \frac{19}{40 - 15.5}$$

$$\alpha = 37.8°$$

Worked example 13

The diagram shows a uniform rectangular lamina that has had a hole cut in it. The centre of mass of the lamina is a distance x from AD and a distance y from AB. Find x and y. The lamina is suspended from the corner A. Find the angle between AB and the vertical.

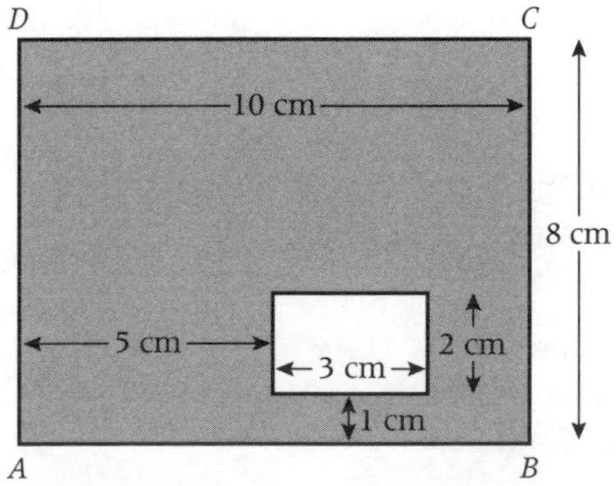

Solution

Consider the table below:

	Area	Distance of C of M from AD	Distance of C of M from AB
Large rectangle	80 cm^2	5 cm	4 cm
Small rectangle	6 cm^2	6.5 cm	2 cm
Lamina	74 cm^2	x	y

Considering the large rectangle as being made up of the lamina and a small rectangle allows us to formulate the equation below, where x is the distance of the centre of mass from AD.

$$80 \times 5 = 74x + 6 \times 6.5$$

$$400 = 74x + 39$$

$$x = \frac{400 - 39}{74}$$

$$= \frac{361}{74} = 4.88 \text{ cm to three significant figures}$$

Similarly to find the distance, y, of the centre of mass from the side AB.

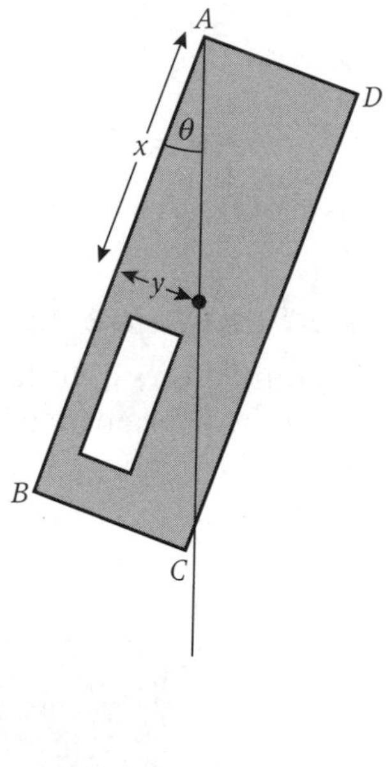

$$80 \times 4 = 74y + 6 \times 2$$

$$320 = 74y + 12$$

$$y = \frac{320 - 12}{74}$$

$$= \frac{308}{74} = 4.16 \text{ cm to three significant figures}$$

When hanging in equilibrium, the angle, θ, between AB and the vertical is:

$$\theta = \tan^{-1}\left(\frac{y}{x}\right)$$

$$= \tan^{-1}\left(\frac{308}{74} \times \frac{74}{361}\right)$$

$$= 40.5°$$

EXERCISE 9D

1 A sphere of mass 2 kg and radius 20 cm is fixed to the end of a rod of length 2 m and mass 4 kg. Find the distance of the centre of mass from the end of the rod.

2 A bat can be modelled as a disc of radius 8 cm and mass 100 g attached to the end of a rod of mass 80 g and length 5 cm. Find the distance between the bottom of the handle and the centre of mass.

3 A rectangular lamina has sides of length 10 cm and 20 cm. A string is attached to one corner of the lamina and it is allowed to hang in equilibrium in a vertical plane. Find the angle between the vertical and the longer side of the rectangle.

4 A rectangular lamina hangs in equilibrium from one corner. The dimensions of the rectangle are 45 cm and 60 cm. Find the angle between the vertical and the shorter sides of the lamina.

5 The diagram shows a uniform lamina.

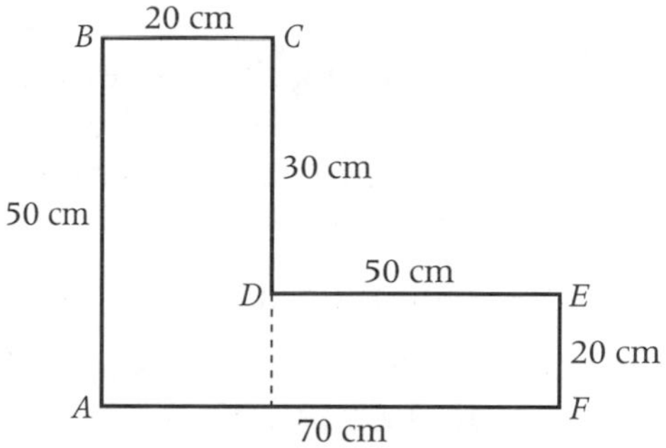

(a) Find the distance between the side *AB* and the centre of mass.

(b) Find the distance between the side *AF* and the centre of mass.

(c) If the lamina is suspended in equilibrium from the corner *A*, find the angle between *AB* and the vertical.

(d) If the lamina is suspended in equilibrium from the corner *B*, find the angle between *AB* and the vertical.

6 A uniform rectangle has sides of length 20 cm and 60 cm. A smaller rectangle, with sides of length 30 cm and 10 cm, is cut out of one corner of the rectangle, to form a lamina. The lamina is then allowed to hang from the opposite corner. When the lamina hangs in equilibrium, what is the angle between the vertical and its longest side?

7 The diagram shows a uniform lamina. Find the distance of the centre of mass from *AB* and from *AF*. The lamina is suspended from the corner *A*. Find the angle between the side *AB* and the vertical, when the lamina is at rest.

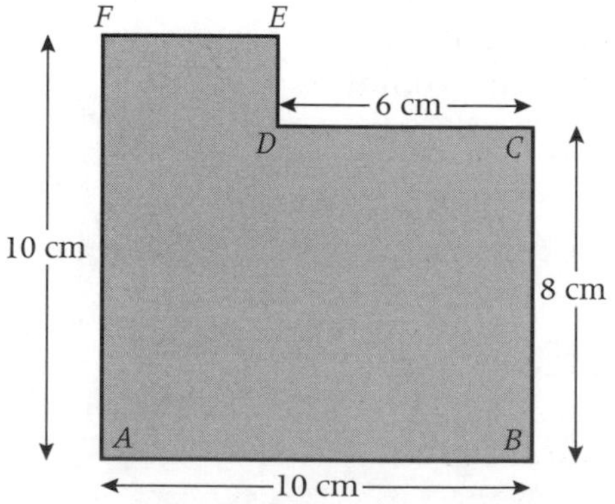

8 A letter F is made out of uniform card, by cutting out two rectangles of size 20 cm by 10 cm and one rectangle of size 40 cm by 10 cm. These are stuck together so that the rectangles overlap as shown in the diagram.

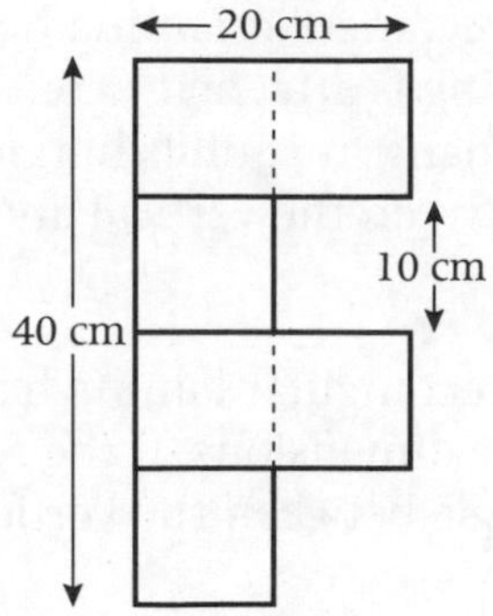

(a) If the F is suspended from the top left-hand corner, find the angle between the vertical and the 40 cm side.

(b) Describe how to suspend the letter, so that the top is horizontal.

9 A light, rectangular framework *ABCD* has particles of mass 3 kg, 4 kg and 3 kg attached at *A*, *B* and *C*, respectively. When suspended from *A* the framework hangs with *C* directly below *A*.

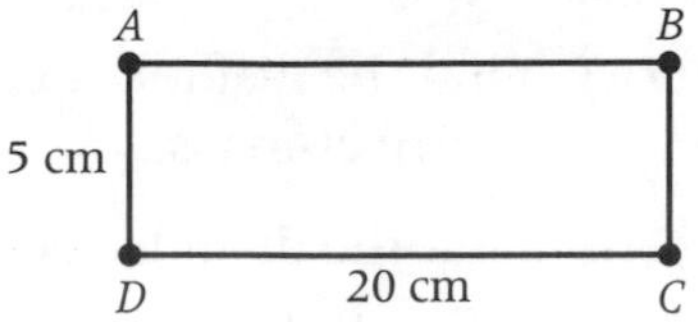

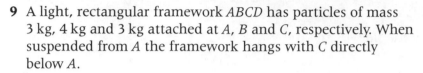

(a) Find the mass of the particle at *D*.

(b) The framework is suspended from a point *X* on *AB*, so that *AB* is horizontal. Find the distance of *X* from *A*.

10 A circular hole, of radius 4 cm, was cut in a uniform, circular disc of radius 10 cm.

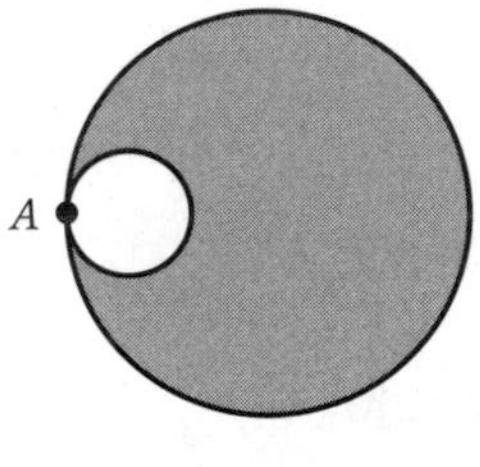

(a) Find the distance of the centre of mass from the point *A*.

(b) If the disc is suspended from *A* and hangs in equilibrium, find the angle between the diameter through *A* and the vertical.

11 A disc, of radius 20 cm, has centre *O*. A hole is radius 5 cm is drilled in the disc, so that the centre of the hole is 5 cm from *O* and lies on the line *BC*, that passes through *O*. The disc is suspended from the point *A*, so that the line *BC* is horizontal. Find the $\angle BOA$.

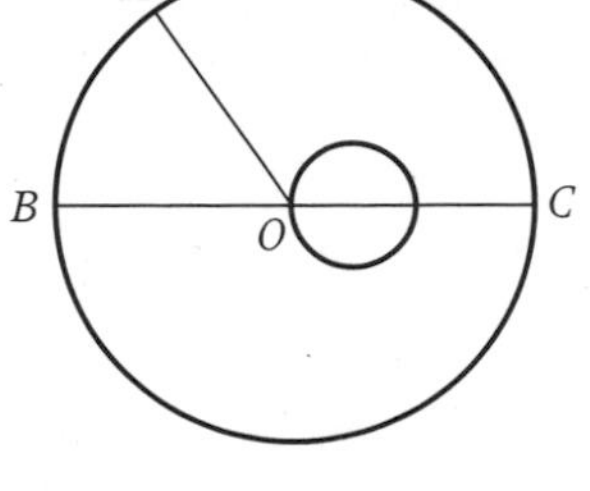

12 The diagram shows a uniform L-shaped lamina $ABCDEF$, of mass $3M$, where $AB = BC = 2a$ and $AF = FE = ED = DC = AH = a$.

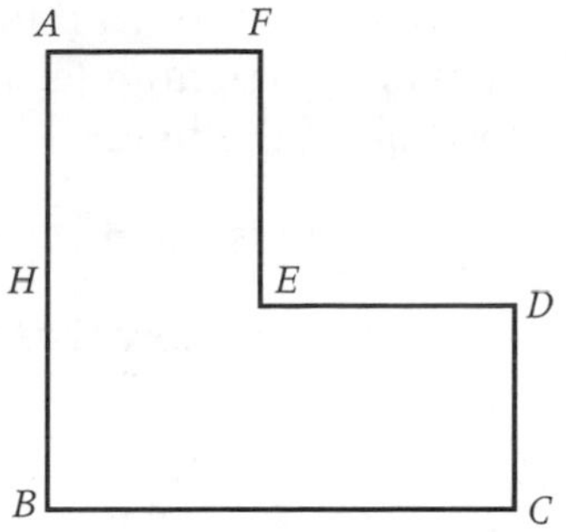

Find the perpendicular distances of the centre of mass of the lamina from the sides AB and BC. The lamina is suspended freely from H, the mid-point of AB, and hangs in equilibrium.

(a) Show that the tangent of the angle which the side AB makes with the horizontal is $\frac{1}{5}$.

(b) When a particle of mass m is attached to F the lamina hangs in equilibrium with AB horizontal. Find m in terms of M.

(c) The particle is now removed and the lamina hangs in equilibrium with BE horizontal when a vertical force of magnitude P is applied at B. Find P in terms of M and g.

13 The diagram shows a road barrier that is made of a metal bar, of mass 50 kg and length 3.6 m, fixed to a counterweight, of mass 150 kg. The barrier is smoothly pivoted at the point O. The unit vectors **i** and **j** are horizontal and vertical, respectively.

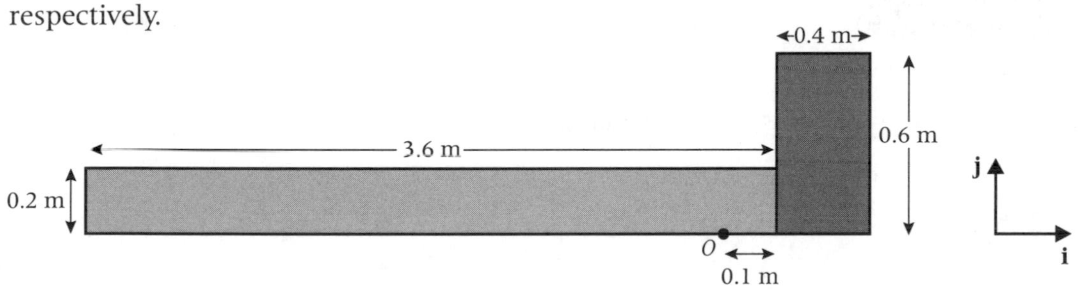

(a) Find the position of the centre of mass of the barrier with respect to the point O in terms of the unit vectors **i** and **j**.

(b) Find the angle between the metal bar and the horizontal when the centre of mass of the barrier is directly above O. [A]

14 The diagram shows four light rods which are rigidly joined together to form a square $OBCD$ of side $2a$. Particles of mass $2m$ are attached to the mid-points of OB, BC and DO, particles of mass $5m$ and m are attached at O and B respectively. The centre of mass of the five particles is $G(\bar{x}, \bar{y})$.

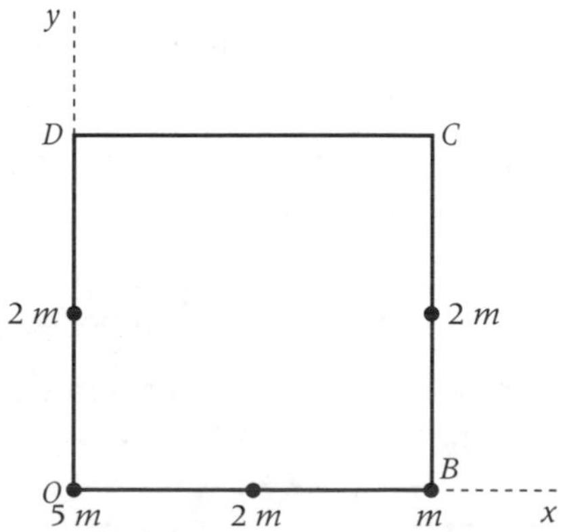

(a) Prove that $\bar{x} = \frac{2}{3}a$.

(b) Find the distance OG.

The system is freely suspended from O and hangs in equilibrium with OB inclined at an angle θ to the downward vertical. Prove that $\tan \theta = 0.5$. [A]

15 A square piece of thin uniform metal, has sides of length 5 cm and a rectangular hole cut in it. The diagram shows the position of the hole.

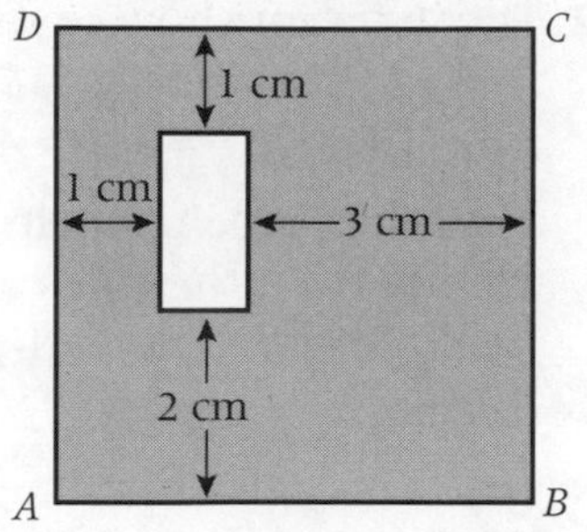

(a) If x cm is the distance of the centre of mass from the side AD, show that $x = \dfrac{119}{46}$.

(b) If y cm is the distance of the centre of mass from the side AB, find y.

(c) The metal sheet is hung from a smooth peg that passes through the rectangular hole. In equilibrium the sheet is at rest, in a vertical plane, with the peg in the corner of the rectangle closest to D. Find the angle between the side AD and the vertical. [A]

Key point summary

Formulae to learn

Moment of a force = Force × Perpendicular distance

Moment of the force about $O = Fd$

Moment of force about $O = Fd \sin \theta$.

$$\bar{x} = \frac{\sum_{i=1}^{n} m_i x_i}{\sum_{i=1}^{n} m_i} \quad \text{and} \quad \bar{y} = \frac{\sum_{i=1}^{n} m_i y_i}{\sum_{i=1}^{n} m_i}$$

- The moment of a force about a point is the magnitude of the force multiplied by the perpendicular from the point to the force. *p124*
- Moments can have a clockwise or anticlockwise turning effect. *p124*
- The resultant force and moment must both be zero for a rigid body to be in equilibrium. *p126*
- The centre of mass of a system of particles or of a composite body can be found using moments. *p136*
- When a body is suspended in equilibrium from a point, the centre of mass is directly below the point of suspension. *p143*

Test yourself

What to review

If your answer is incorrect – review

1 A metal beam, of mass 6 kg and length 2 m, rests in a horizontal position on two supports that are at a distance of 40 cm from each end of the beam. A 1.2 kg mass is placed at one end of the beam. — *Section 8.3*

(a) Find the magnitude of the reaction forces acting on the beam.

(b) What is the greatest mass that could be placed at the other end of the beam, if it is to remain in equilibrium?

---

2 A ladder, of mass 20 kg and length 5 m, has its base on rough, horizontal ground and rests against a smooth vertical wall. The coefficient of friction between the ground and the ladder is 0.6. The angle between the ladder and the ground is θ. — *Section 8.3*

(a) Find the magnitude of the forces acting on the base of the ladder in terms of g and θ.

(b) Find the smallest value of θ for which the ladder will remain at rest.

---

3 The diagram shows a uniform lamina *ABCDEF*. — *Section 8.3*

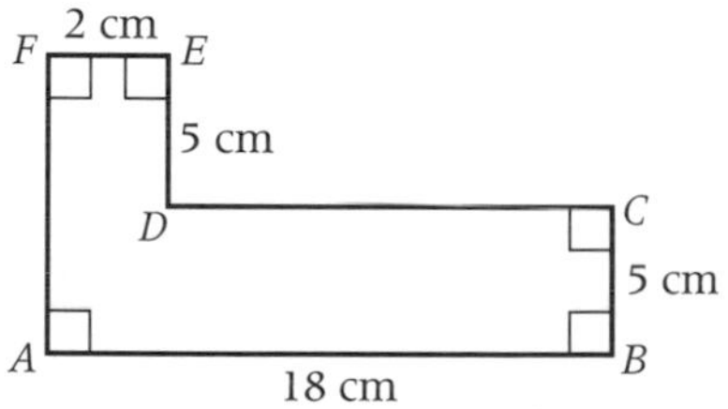

(a) Show that the centre of mass of the lamina is 8.2 cm from *AF*.

(b) Find the distance of the centre of mass from *AB*.

(c) If the lamina is suspended from the corner *A*, find the angle between *AB* and the vertical.

Test yourself ANSWERS

1 (a) 25.5 N, 45.1 N. **(b)** 13.8 kg.

2 (a) $R = 20g$, $F = \frac{10g}{\tan\theta}$ **(b)** 39.8°.

3 (b) 3 cm. **(c)** 20.1°.

9

Exam style practice paper

Time allowed 1 hour 45 minutes

Answer **all** questions

---

1 A car, of mass 1200 kg, accelerates from rest to a speed of $10\ \text{m s}^{-1}$, as it travels a distance of 40 m. Assume that the acceleration of the car is constant.

(a) Calculate the acceleration of the car and the magnitude of the resultant force on the car. *(4 marks)*

(b) If the car continues with the same acceleration, how long would it take to reach a speed of $20\ \text{m s}^{-1}$ and how far would it have travelled since setting off. *(4 marks)*

(c) Explain why it is unlikely that the car would continue to move with the same acceleration. *(2 marks)*

2 A light rod has length 2 m. Particles of mass 8 kg, 5 kg and 7 kg are fixed to the rod as shown in the diagram below..

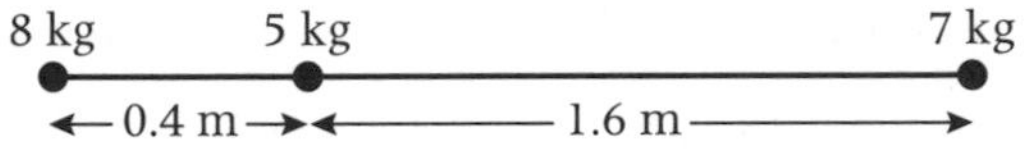

(a) Find the distance of the centre of mass of the rod from the left-hand end. *(3 marks)*

(b) The rod is placed so that it rests in equilibrium, in a horizontal position, with the 5 kg particle and the 7 kg particle resting on supports.

(i) Determine the magnitude of the upward force exerted by each support. *(4 marks)*

(ii) An additional particle of mass m kg is attached to the 8 kg mass. Determine the maximum possible value of m for which the rod remains in equilibrium. *(2 marks)*

3 The diagram shows two particles of mass 3 kg and 7 kg that are joined by a light inextensible string that passes over a light, smooth pulley.

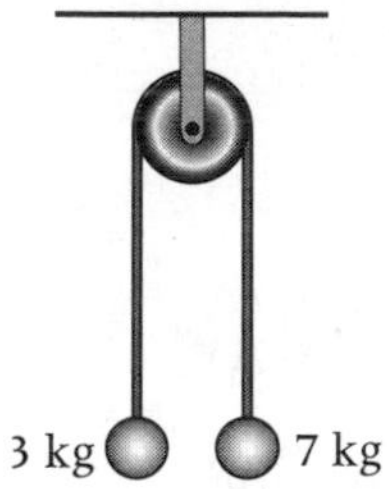

The particles are initially at the same level.

(a) Find the acceleration of the particles. *(4 marks)*

(b) Find the tension in the string. *(2 marks)*

(c) Find the speed of the particles when they are 20 cm apart. *(2 marks)*

4 A car, of mass 1000 kg, skids for 20 m and then hits a car of mass 1200 kg that is moving at 10 m s^{-1} in the same direction. After the collision, both cars move together at a speed of 8 m s^{-1}. The coefficient of friction between the tyres and the rod for the first car is 0.8. Assume that all the motion takes place on a horizontal surface.

(a) Use conservation of momentum to find the speed of the first car, just before the collision. *(2 marks)*

(b) Find the speed of the first car when it begins to skid. *(4 marks)*

5 A child pulls a sledge, of mass 20 kg, along a snow-covered surface. The child uses a rope that makes an angle of 30° with the ground as he pulls. The coefficient of friction between the ground and the sledge is 0.2. The tension in the rope is T N.

(a) Find, in terms of T, the magnitude of the normal reaction force acting on the sledge, when it is pulled on a horizontal surface. *(3 marks)*

(b) Find the tension in the rope if the sledge accelerates at 0.05 m s^{-2} on a horizontal surface. *(4 marks)*

(c) Find the tension needed to keep the sledge moving at a constant speed. *(2 marks)*

(d) The child then pulls the sledge up a slope inclined at an angle of 10° to the horizontal. The rope makes an angle of 30° with the slope. Find the tension if the sledge moves at a constant speed. *(4 marks)*

6 A helicopter is initially at rest and has position vector $(30\mathbf{i} + 400\mathbf{j} + 120\mathbf{k})$ m, where the unit vectors **i**, **j** and **k** are east, north and vertical respectively. The helicopter moves with a constant acceleration so that 10 seconds later its position is $(100\mathbf{i} + 450\mathbf{j} + 140\mathbf{k})$ m.

(a) Find the velocity of the helicopter after 10 seconds. *(4 marks)*

(b) Find an expression for the position of the helicopter at time t seconds. *(3 marks)*

After the first 10 seconds the helicopter stops accelerating and moves with a constant velocity.

(c) Find the position vector of the helicopter when it reaches a height of 200 m. *(4 marks)*

7 A shot is thrown with an initial velocity $V\,\text{m s}^{-1}$, at an angle α above the horizontal.

(a) A simple model assumes that the height of release of the shot is zero. Find an expression for the range of the shot in terms of V, g and α. *(6 marks)*

(b) Calculate the range if $V = 10\,\text{m s}^{-1}$ and $\alpha = 40°$. *(3 marks)*

In fact the shot is released at a height of 1.8 m with the values of V and α given in part **(b)**.

(c) What would happen if the shot was thrown in this way inside a hall of length 10 m and height 3.5 m? *(4 marks)*

8 A ladder, of mass 12 kg and length 5 m, rests in equilibrium with its base on a rough horizontal surface and the other end resting against a smooth vertical wall. The ladder is at an angle of 60° to the horizontal. The coefficient of friction between the ladder and the ground is μ. A man, of mass 70 kg, climbs to the top of the ladder.

(a) Find the magnitude of the forces acting on the base of the ladder, when the man is at the top. *(5 marks)*

(b) Find an inequality that μ must satisfy if the ladder is to remain at rest. *(2 marks)*

Answers

EXERCISE 2A

1 84 m, 1.6 m s^{-2}, 0 m s^{-2}, -2 m s^{-2}.

2 104 m.

5 **(a)** 0 m s^{-2}, **(b)** 2 m s^{-2}, **(c)** $-\frac{5}{3}$ m s^{-2}, **(d)** 437.5 m.

6 584 m.

7 0.25 m s^{-2}, 0 m s^{-2}, -0.5 m s^{-2}, 8000 m.

8 **(a)** 16 000 m, **(b)** $233\frac{1}{3}$ s.

9 1.2 m s^{-2}, -1.8 m s^{-2}, 85 s, 1305 m.

10 9200 m, 440 s.

11 **(a)** 10 m s^{-1}, 2.5 m s^{-2}, **(b)** $\frac{16}{9}$ s, 5.06 m s^{-2}.

12 **(a)** 40 m, 20 m s^{-1}, **(b)** 18 s, 6.67 m s^{-2}.

EXERCISE 2B

1 **(a)** 100 m, 20 m s^{-1}, **(b)** 22.5 m s^{-1}, 206.25 m.

2 **(a)** 0.3 m s^{-2}, **(b)** 33.3 s.

3 **(a)** 0.7 m s^{-2}, **(b)** 85 m.

4 **(a)** 0.2 m s^{-2}, **(b)** 10 m, **(c)** 15 m.

5 0.417 m s^{-2}.

6 361 m.

7 -10 m s^{-2}, 1.69 s.

8 **(a)** 889 m, **(b)** 66.7 s.

9 1200 m.

10 **(a)** 2.92 m s^{-2}, 18.7 m s^{-1}, **(b)** 8.55 s.

11 **(a)** 16.3 m s^{-1}, **(b)** 2.21 m s^{-2}.

12 $0.580\ \text{m s}^{-2}$, $15.3\ \text{m s}^{-1}$.

13 9.33 s.

14 12 400 m.

15 **(a)** $24\ \text{m s}^{-1}$. **(b)** 7800 m.

16 8 m, 7.5 s.

17 **(a)** 25 m, **(b)** 20 s, **(c)** $50\ \text{m s}^{-1}$. $62.5\ \text{m s}^{-1}$.

18 **(a)** $t = 4$ s, **(b)** no, as vehicles are 36 m apart.

19 **(a)** 36 m, **(b)** 36 m + car length.

EXERCISE 2C

1 **(a)** 0.639 s, **(b)** 1.58 s.

2 **(a)** 0.782 s, **(b)** $7.67\ \text{m s}^{-1}$.

3 **(a)** 1.5 s, **(b)** 12.0 m, **(c)** $15.4\ \text{m s}^{-1}$.

4 **(a)** $2\ \text{m s}^{-2}$, $20\ \text{m s}^{-1}$, **(b)** 120.4 m.

6 **(a)** 2.02 s, **(b)** 2.02 s.

7 2.10 s, $24.6\ \text{m s}^{-1}$.

8 $6.64\ \text{m s}^{-1}$, $-40.1\ \text{m s}^{-1}$.

9 **(a)** The one thrown down, **(b)** both have the same speed, **(c)** 5 m, $5.41\ \text{m s}^{-1}$.

10 $32.75\ \text{m s}^{-1}$.

11 $\frac{h}{4}$.

12 **(a)** 12 m, constant acceleration, **(b)** 31.1 m.

13 **(a)** 0.569 m, **(b)** over-estimate due to air resistance.

EXERCISE 3A

1 **(a)** $0\mathbf{i} + 0\mathbf{j}$, $30\mathbf{i} + 20.1\mathbf{j}$, $60\mathbf{i} + 30.4\mathbf{j}$, $90\mathbf{i} + 30.9\mathbf{j}$, $120\mathbf{i} + 21.6\mathbf{j}$, $150\mathbf{i} + 2.5\mathbf{j}$, $180\mathbf{i} - 26.4\mathbf{j}$, **(c)** 153 m.

2 **(a)** $0\mathbf{i} + 1\mathbf{j}$, $1\mathbf{i} + 1.6\mathbf{j}$, $2\mathbf{i} + 1.8\mathbf{j}$, $3\mathbf{i} + 1.6\mathbf{j}$, $5\mathbf{i} + 0\mathbf{j}$.

3 Children collide.

4 **(a)** $6\mathbf{i} - 9\mathbf{j}$, $86\mathbf{i} + 111\mathbf{j}$, $166\mathbf{i} + 231\mathbf{j}$, $246\mathbf{i} + 351\mathbf{j}$, **(c)** 11.1 m, **(d)** 90 s, 96.7 s.

5 **(a)** 1.225 m, **(b)** 0.5 s, **(c)** 90 m.

6 $16\mathbf{i} + 8\mathbf{j} + 0\mathbf{k}$.

EXERCISE 3B

1 **(a)** $45\cos 10°\mathbf{i} + 45\sin 10°\mathbf{j} = 44.3\mathbf{i} + 44.3\mathbf{j}$,
(b) $105\cos 30°\mathbf{i} + 105\sin 30°\mathbf{j} = 90.9\mathbf{i} + 52.5\mathbf{j}$,
(c) $-21\cos 70°\mathbf{i} + 21\sin 70°\mathbf{j} = -7.18\mathbf{i} + 19.7\mathbf{j}$,
(d) $-62\cos 10°\mathbf{i} - 62\sin 10°\mathbf{j} = -61.1\mathbf{i} - 10.8\mathbf{j}$,
(e) $290\cos 72°\mathbf{i} - 290\sin 72°\mathbf{j} = 89.6\mathbf{i} - 4.24\mathbf{j}$.

2 $-6\cos 45°\mathbf{i} - 6\sin 45°\mathbf{j} = -4.24\mathbf{i} - 4.24\mathbf{j}$.

3 **(a)** $5\mathbf{i}$, **(b)** $-5\mathbf{j}$, **(c)** $-5\mathbf{i}$, **(d)** $5\cos 45°\mathbf{i} - 5\sin 45°\mathbf{j}$,
(e) $5\cos 45°\mathbf{i} + 5\sin 45°\mathbf{j} = 3.54\mathbf{i} + 3.54\mathbf{j}$.

4 $8\cos 50°\mathbf{i} + 8\sin 50°\mathbf{j} = 5.14\mathbf{i} + 6.13\mathbf{j}$.

5 **(a)** 8.06 m s^{-1}, 029.7°, **(b)** 7.81 m s^{-1}, 140.2°, **(c)** 12.0 m s^{-1}, 221.6°,
(d) 14.4 m s^{-1}, 303.7°.

6 **(a)** 8.54 m, **(b)** 159.4°.

7 3.61 m s^{-2}, 303.7°.

EXERCISE 3C

1 $-0.3\mathbf{i} - 0.5\mathbf{j}$.

2 **(a)** $4.8\mathbf{i} + 2.4\mathbf{j}$, **(b)** $31.7\mathbf{i} + 8.6\mathbf{j}$.

3 $\mathbf{v} = (3 + t)\mathbf{i} + (-5 + t)\mathbf{j}$, $\mathbf{r} = \left(3t + \dfrac{t^2}{2}\right)\mathbf{i} + \left(-5t + \dfrac{t^2}{2}\right)\mathbf{j}$.

4 $36\mathbf{i} + 48\mathbf{j}$, $108\mathbf{i} + 144\mathbf{j}$.

5 $\mathbf{r} = 4t\mathbf{i} + (2 + 9t - 5t^2)\mathbf{j}$, $8\mathbf{i}$.

6 $r = (0.2 + \sqrt{3}t)\mathbf{i} + (0.1 + t)\mathbf{j}$, $1.76\mathbf{i} + 1\mathbf{j}$.

7 **(b)** 6.59 s, $r = 46.1\mathbf{i}$, **(c)** 54.1 m, **(d)** 33.3 m s^{-1}.

8 **(a)** 6 s, $120\mathbf{i}$, **(b)** 4 s.

9 **(a)** $\mathbf{v} = (4 - 0.02t)\mathbf{i} + (6 - 0.04t)\mathbf{j}$, $\mathbf{r} = (80 + 4t - 0.01t^2)\mathbf{i} +$
$(20 + 6t - 0.02t^2)\mathbf{j}$, **(b)** $416\mathbf{i} + 92\mathbf{j}$, **(c)** $0.4\mathbf{i} - 1.2\mathbf{j}$.

10 **(a)** $\mathbf{r}_H = 6t\mathbf{i} + (80 - 3t)\mathbf{j}$, $\mathbf{r}_A = (5t + 0.05t^2)\mathbf{i} + (10 + 0.025t^2)\mathbf{j}$,
(b) 20 s, $120\mathbf{i} + 20\mathbf{j}$.

11 $\mathbf{v} = (2 - 0.06t)\mathbf{i} + (3 - 0.04t)\mathbf{j}$, $\mathbf{r} = (40 + 2t - 0.03t^2)\mathbf{i} +$
$(20 + 3t - 0.02t^2)\mathbf{j}$, (b) 60 s.

12 **(a) (i)** 500 s, **(ii)** $10\,000\mathbf{i}$, **(iii)** 100.5 m s^{-2}.

EXERCISE 4B

1 **(a)** 5 N, 36.9°, **(b)** 14 N, 38.2°, **(c)** 12.8 N, 43.0°, **(d)** 4.35 N, 151.4°,
(e) 3.78 N, 84.9°.

2 **(a)** 90°, **(b)** 91.5°, **(c)** 152.3°, **(d)** 158.2°.

3 11 N, 1 N.

4 15.8 N.

5 11.1 N.

6 158.2°.

7 7.43 N, 53.8°.

EXERCISE 4C

1 **(a)** 1 N, **(b)** 5.96 N.

2 5.92 N.

3 $a = -7$, $b = -5$.

4 **(a)** $9\mathbf{i} + 17\mathbf{j}$, **(b)** 19.2 N, **(c)** 62.1°.

5 3.61 N, 123.7°.

6 $(\lambda - 4)\mathbf{i} + 4\mathbf{j}$, $\sqrt{\lambda^2 - 8\lambda + 32}$, $\lambda = 1$ or 7, 53.1°.

EXERCISE 4D

1 **(a)** 4.50, 5.36, **(b)** −7.25, 3.38, **(c)** −5.79, −6.89.

2 **(a)** $4.50\mathbf{i} + 5.36\mathbf{j}$, **(b)** $-7.25\mathbf{i} + 3.38\mathbf{j}$, **(c)** $-5.79\mathbf{i} - 6.89\mathbf{j}$.

3 **(a)** $-\frac{Mg}{2}$, $-\frac{Mg\sqrt{3}}{2}$, **(b)** $-W\sin\alpha$, $-W\cos\alpha$.

4 **(a)** 5.92 N, 88.5° below x, **(b)** 1.43 N, 82.3° below x, **(c)** 7.21 N, 46.3° below x, **(d)** 4.56 N, 70.3° below x.

5 0 N.

6 **(a)** $-\frac{W}{\sqrt{2}}$, $-\frac{W}{\sqrt{2}}$, **(b)** $-\frac{W}{2}$, $-\frac{W\sqrt{3}}{2}$, **(c)** $-\frac{W\sqrt{3}}{2}$, $\frac{W}{2}$.

7 $-mg\sin\alpha$.

EXERCISE 4E

1 **(a)** $F_1 = 1.71$ N, $F_2 = 4.70$ N, **(b)** $F_1 = 4.01$ N, $F_2 = 17.7$ N, **(c)** $F_1 = 5.22$ N, $F_2 = 7.04$ N.

2 **(a)** $F = 25$ N, $\theta = 73.7°$, **(b)** $F = 9.42$ N, $\theta = 15.8°$, **(c)** $F = 8.64$ N, $\theta = 130.4°$.

3 **(a)** $F_1 = 7.83$ N, $F_2 = 6.21$ N, **(b)** $F_1 = 6.77$ N, $F_2 = 9.44$ N, **(c)** $F_1 = 6.72$ N, $F_2 = 16.4$ N, **(d)** $F_1 = 10$ N, $F_2 = 3.34$ N, **(e)** $F_1 = 4.79$ N, $F_2 = 6.16$ N, **(f)** $F_1 = 131$ N, $F_2 = 176$ N.

4 **(a)** $-8\mathbf{i} - 4\mathbf{j}$, **(b)** 8.94 N, 153.4° below $\mathbf{i}$.

5 $a = -11$, $b = 0$.

6 **(a)** $\mathbf{F}_2 = -8\mathbf{i} - 9\mathbf{j}$, **(b)** $\mathbf{F}_3 = 2\mathbf{i} - 2\mathbf{j}$.

7 None.

8 24.5 N, 42.4 N.

9 38.7 N, 72.7 N.

10 5.57 N, 17.9°.

11 $P = 8.39$ N, $R = 13.1$ N.

12 11.2 N, 26.6°.

13 124.2°, 97.2°, 138.6°.

14 1.81 kg.

15 27.1 N, 24.0 N.

16 $-90\mathbf{i} + 40\mathbf{j} + 132\mathbf{k}$, 165 N.

17 $-88\mathbf{i} - 128\mathbf{j}$, 155 N.

EXERCISE 4F

1 0.255.

2 **(a)** No motion, **(b)** slides, **(c)** no motion.

3 0.567.

4 73.6 N.

5 51.8 N, 185°, 0.529.

6 0.5.

7 88.2 N.

8 **(a)** 0.085, **(b)** 30.8 N.

EXERCISE 4G

1 **(a)** Slides, **(b)** no motion, **(c)** slides.

2 **(a)** 7.37 N, **(b)** 24.3 N, **(c)** 44.2 N.

3 **(a)** 26.7 N, **(b)** 42.7 N, **(c)** 70.4 N.

4 34.6 N.

5 **(a)** 28.3 N, **(b)** 181 N, **(c)** 671 N.

6 **(a)** 29.2 N, **(b)** 7.88 N, **(c)** 14.5 N.

7 $F = 21.9$ N, $R = 77.9$ N, 0.281.

8 **(a)** $R = 53.9$ N, $F = 33.6$ N, $\mu \geq 0.623$,
(b) $R = 36.8$ N, $F = 6.59$ N, $\mu \geq 0.179$,
(c) $R = 36.8$ N, $F = 38.4$ N, $\mu \geq 1.04$.

EXERCISE 5A

1 **(a)** $X = 6$ N, $Y = 5$ N, **(b)** $X = 10$ N, $Y = 12$ N, **(c)** $X = 69.3$ N, $Y = 69.3$ N.

2 **(a)** 10 N, 36.9°, **(b)** 15.7 N, 60°, **(c)** 8.66 N, 60°.

3 1350 N.

4 0.245.

5 **(a)** $T = 115$ N, **(b)** $R = 112$ N.

6 210 N.

7 49.5 m s^{-1}.

8 17.9 m s^{-1}.

9 43.39 m s^{-1}.

EXERCISE 5B

1 0.5 m s^{-2}.

2 6 N.

3 $(2.5\mathbf{i} + \mathbf{j})$ m s^{-2}.

4 0.7 m s^{-2}.

5 **(a)** 2970 N, **(b)** 2880 N, **(c)** 2910 N.

6 **(a)** $a = 8$ m s^{-2}, **(b)** $a = 0.8$ m s^{-2}, $R = 98$ N, **(c)** $T = 47.3$ N, $R = 74.3$ N, **(d)** 103 N, **(e)** $\theta = 30°$, $a = 0.329$ m s^{-2}, **(f)** $T = 66.5$ N, $a = 3$ m s^{-2}.

7 4.77 m s^{-2}.

8 87.1°.

9 4167 N.

10 4.9 m s^{-2}.

11 4 s.

12 33.3 N, 0.195.

13 11.8°.

14 61 500 N, 49 000 N, 24 000 N, 10.3 m.

15 $P = 10$ N.

16 $(13\mathbf{i} + 20\mathbf{j})$ m.

17 **(a)** 12.1°, **(b)** 116 000 N, 0.0058 m s^{-2}.

18 **(a)** 253 N, **(b)** 10.2 N, **(c)** 0.340 m s^{-2}.

19 **(a)** 8 m, 7.5 s, **(b)** first stage 4240 N.

20 **(b)** 2428 N, **(c)** will stop accelerating.

21 **(a)** $\frac{m(a + g\sqrt{3})}{2}$, **(b)** 17.0 m s^{-2}, 5.66 m s^{-2}, unlikely but possible, **(c)** no.

22 **(a)** 16 000 m, **(b)** $233\frac{1}{3}$ s, **(c)** 10 000 N in opposite direction to motion.

23 **(a)** 194 N, **(b)** 3.24 m s^{-2}.

EXERCISE 5C

4 $T = 2863$ N, $R = 803$ N, $W = 1960$ N.

EXERCISE 6A

1 2.45 m s^{-2}, 36.75 N.

2 **(a)** 1.09 m s^{-2}, 43.6 N, **(b)** 1.96 m s^{-2}, 47.0 N, **(c)** 3.77 m s^{-2}, 54.3 N.

3 **(a)** 2.67 m s^{-2}, **(b)** 0.612 s.

4 1.96 m s^{-2}, 11.76m N.

5 0.25.

6 0.98 m s^{-2}, 17.6 N.

7 **(a)** 2.43 m s^{-2}, **(b)** 22.1 N, **(c)** −2 m s^{-2}.

8 **(a)** 1100 N, **(b)** 400 N.

9 1.93 m s^{-2}, 15.7 N, 2.78 m s^{-1}.

10 **(b)** $\frac{mMg(1 + \mu)}{m + M}$, **(c)** $\sqrt{\frac{m + M}{g(m - \mu M)}}$, $\sqrt{\frac{g(m - \mu M)}{m + M}}$.

11 **(a)** 3.92 m s^{-2}, **(b)** 27.6 N.

12 **(a)** 0.668 m s^{-2}, **(b)** 9.48 s.

13 **(a)** 0.942 m s^{-2}, **(b)** 0.250 m s^{-2}.

EXERCISE 7A

1 **(a)** 2.60 s, **(b)** 33.1 m, **(c)** 8.27 m.

2 35.3 m.

3 3.18 m.

4 15.3 m, 17.7 m.

5 12.0 m, 4.11 m.

6 2.70 m s^{-1}, 0.250 m, air resistance would reduce the range.

7 100 m, 4.94 s, 98.8 m.

8 3.05 s, 52.8 m.

9 Yes, predicts 108 m, unlikely to cause a problem.

10 Passes over the wall at a height of 3.02 m.

11 $18\frac{1}{8}$ m, 21.7 m.

12 1.84 m, 1.22 s, 12.7 m, 10.5 m s^{-1}, 6.0, 11.1 m s^{-1}, −20.1°.

13 23.6 m s^{-1}.

14 25.3 m s^{-1}, clears net by 31.6 cm.

15 6.64 m.

16 2.63 m, 2.77 m.

17 **(a)** No air resistance, horizontal ground, **(b)** 105 m, **(c)** increase it.

18 **(a)** **(ii)** 23.2 m s^{-1}, **(b)** yes.

EXERCISE 7B

1 23.7° or 66.4°.

2 4.7° or 85.3°, 4.7°.

4 5.3° or 84.7°, 0.753 s, 8.12 s, 163 m.

5 29.7 m s^{-1}, 4.29 s.

6 31.7 m s^{-1}, 38.7°.

7 $\frac{4a}{T}, \frac{gT}{2}$.

8 76.0°.

9 **(a)** $u\cos\alpha$, $u\sin\alpha - gt$, **(b)** $u\cos\alpha t$, $u\sin\alpha t - \frac{1}{2}gt^2$, $\frac{gT^2}{8}$.

10 **(a)** $\frac{2UV}{g}$, **(b)** **(i)** ×2, **(ii)** ×4, **(c)** 2.89 s.

11 **(a)** horizontal ground and no forces other than gravity, **(b)** 19.1 m, **(c)** 17.2 m, as 2 m would be a big difference.

12 **(b)** 8.84 m, **(c)** impossible as max range is 91.8 m.

EXERCISE 7C

1 8.43 m.

2 Yes for angles between 3.2° below horizontal and 47.1° above the horizontal.

3 No, $62.4° < \theta < 72.6°$.

4 63.4° or 27.6°.

5 28 m s^{-1}, 26.6°.

6 3.57 s.

7 **(b)** Size of ball and air resistance, **(c)** 27.34° to 74°, **(d)** hit the bar.

8 32.8 m s^{-1}.

9 **(b)** 7.23 m s^{-1}, **(c)** 68.6° or 35.4°.

11 $a = 82.5$, $b = 1.75$, **(a)** 1.75, **(b)** 0.5 or 3.

12 74.9°–75.3°.

EXERCISE 8A

1 **(a)** 4 800 000 N s, **(b)** 0.012 N s, **(c)** 12 000 N s.

2 4 m s^{-1}.

3 2.75 m s^{-1}.

4 4 m s^{-1}.

5 2 m s^{-1}.

6 0.618 m s^{-1}.

7 2.14 m s^{-1}.

8 0.05 m s^{-1}.

9 $2u$ in the opposite direction.

10 **(a)** 4.90 m s^{-1}, **(b)** 14.7 m s^{-1}.

11 **(a)** 48.9 kph or 13.6 m s^{-1}, **(b)** 2.31 m s^{-2}, 415 000 N.

12 **(a)** 1 m s^{-1}, **(b)** 8.6 cm.

13 6.33 m s^{-1}.

14 **(a)** $t = 4$ s, **(b)** no, as 36 m is the distance between the cars, **(c)** 7.2 m s^{-1}.

15 **(a)** Speed and size, **(c)** 580 N, **(d)** $k = 4.41$, **(e)** 6.92 m s^{-1}.

16 **(c)** 28.0 m s^{-1}, **(d)** 800 kg.

EXERCISE 9A

1 **(a)** 40 N m, **(b)** −300 N m, **(c)** 240 N m, **(d)** 21.2 N m, **(e)** −20.8 N m, **(f)** −68.9 N m, **(g)** −7.62 N m, **(h)** −23.1 N m, **(i)** −17.3 N m.

2 **(a)** 0.5 m, **(b)** 0.943 m, **(c)** 1.15 m.

EXERCISE 9B

1 1.5 m from centre.

2 270 N, 515 N.

3 **(a)** 196 N each, **(b)** 294 N, 392 N.

4 **(a)** 110 N, 36.8 N, **(b)** 7.5 kg.

5 **(a)** 32.7 N, 163 N, **(b)** 4 kg.

6 1.05 kg.

7 Within 29 cm of the centre.

8 120 g, 360 g.

9 **(a)** B, **(b)** 700 g.

10 **(a)** 20 N m.

11 86.6 N.

12 **(a)** 438 N, **(b)** 314 N.

13 **(a)** 35.5 N, **(b)** 69.5 N.

14 56.6 N, 196 N.

16 **(c)** Up to 3 m, **(d)** **(ii)** anybody can.

17 **(c)** 594 N, **(d)** 343 N.

18 **(b)** 196 N, **(c)** yes, as μ only needs to be small, **(d)** no, as moment of tension is zero.

EXERCISE 9C

1 1.2 m.

2 0.45 m.

3 2.06 m.

4 1.25 m, 8 kg.

5 **(a)** 0.1 m, **(b)** 0.05 m.

6 **(a)** (0.467, 0.167), **(b)** (0.13, −0.05), **(c)** (0.985, 0.368), **(d)** (0.2, −0.267).

7 **(a)** $(0.7l, 0.5l)$, **(b)** $(0.643l, 0.5l)$.

8 $m = 6$ kg, $M = 4$ kg.

9 **(a)** 2.5 kg, **(b)** 10.7 cm.

10 4.95 cm.

EXERCISE 9D

1 1.4 m.

2 8.33 cm.

3 26.6°.

4 53.1°.

5 **(a)** 25 cm, **(b)** 17.5 cm, **(c)** 55°, **(d)** 37.6°.

6 18.4°.

7 4.45 cm, 4.73 cm, 43.3°.

8 **(a)** 23.2°, **(b)** on top edge, 7.5 cm from top left corner.

9 **(a)** 4 kg, **(b)** 10 cm.

10 11.1 cm.

11 $\angle BOA = 81.4°$.

12 $\frac{5a}{6}, \frac{5a}{6}$, **(b)** $m = \frac{M}{2}$, **(c)** $P = \frac{5Mg}{2}$.

13 **(a)** $-0.2\mathbf{i} + 0.25\mathbf{j}$, **(b)** 38.7°.

14 **(b)** $\frac{a\sqrt{5}}{3}$.

15 **(b)** $\frac{113}{46}$, **(c)** 45.8°.

ANSWERS TO PRACTICE PAPER

1 **(a)** 1.25 m s^{-2},1500 N, **(b)** 16 s, 160 m, **(c)** air resistance would increase with speed and decrease the resultant force and acceleration.

2 **(a)** 0.8 m, **(b) (i)** 147 N, 49 N, **(ii)** 20 kg.

3 **(a)** 3.92 m s^{-2}, **(b)** 41.16 N, **(c)** 0.885 m s^{-1}.

4 **(a)** 17.6 m s^{-1}, **(b)** 25.0 m s^{-1}.

5 **(a)** $R = 196 - \frac{T}{2}$, **(b)** 41.6 N, **(c)** 40.6 N, **(d)** 75.2 N.

6 **(a)** $14\mathbf{i} + \mathbf{j} + 4\mathbf{k}$,
(b) $\mathbf{r} = (30 + 0.7t^2)\mathbf{i} + (400 + 0.5t^2)\mathbf{j} + (120 + 0.2t^2)\mathbf{k}$,
(c) $\mathbf{r} = 310\mathbf{i} + 600\mathbf{j} + 200\mathbf{k}$.

7 **(a)** $\frac{V^2 \sin 2\alpha}{g}$, **(b)** 10.0 m, **(c)** hits the roof.

8 **(a)** $F = 430$ N, $R = 804$ N, **(b)** $\mu \geq 0.535$.

Index

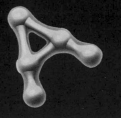